Asset and Liability Portfolio of Farmers

Published by Academic Foundation
in association with

INDIRA GANDHI INSTITUTE OF DEVELOPMENT RESEARCH (IGIDR),
Mumbai

About the Authors

S. Bisaliah, PhD in Agriculture Economics from University of Minnesota, has been engaged in teaching, research, development work and academic administration for the last four decades. He had served as Chairman, Karnataka State Agricultural Price Commission, contributing for policy and programme formulation for agriculture development. The international working experiences of Prof Bisaliah include: Monitoring and Evaluation Expert, World Bank Forestry Project (Nigeria), thrice consultant to FAO (Rome), twice consultant to Economic and Social Commission for Asia and Pacific, and once Consultant to World Bank-FAO. He has 170 research articles and policy reports on agriculture/rural development and capital formation in agriculture.

S. Mahendra Dev is Director, Indira Gandhi Institute of Development Research (IGIDR), Mumbai, India. Earlier he was Chairman of the Commission for Agricultural Costs and Prices, Govt. of India, Delhi and Director, Centre for Economic and Social Studies, Hyderabad (1999 to 2008). He received his PhD from the Delhi School of Economics and completed his Post-doctoral research at Yale. He has more than 100 research publications in national and international journals and has written extensively on agricultural development, poverty and public policy, food security, employment guarantee schemes, social security, farm and non-farm employment. He has been a consultant and adviser to many international organisations, such as UNDP, World Bank, International Food Policy Research Institute, ILO, FAO and ESCAP. He has been a member of several government committees including the Prime Minister's Task Force on Employment and Rangarajan Commission on Financial Inclusion, and is presently member of the expert group on poverty.

Saifullah Syed, presently working in the Policy Assistance and Support Service of the Food and Agriculture Organization (FAO) of the United Nations, holding the position of Senior Policy Officer, joined FAO in 1983. Prior to that, he was working for the Organization for Economic Cooperation and Development (OECD). He completed his MSc in Economics from Birkbeck College, University of London and PhD in Economics from University of Paris-Dauphine, France. His field of specialisation includes: modelling for policy analysis, providing policy assistance and advice for sustainable agriculture and rural development and international trade. Dr Syed has published several research papers and is co-author of the book: *Agriculture in the Cook Islands: New Directions* (1993). His current research interest is in policies for promoting investment in agriculture. He is leading a Japanese Trust Fund Project on policies for promoting investment in agriculture, which led to initiation of work for this book.

Dhritisree Sarkar, is a PhD student at the Indira Gandhi Institute of Development Research (IGIDR), Mumbai. Her research topic is "Feminisation of Poverty in India". She has done her Masters in Economics from Jawaharlal Nehru University, New Delhi. Her research interests include development economics, agricultural economics and econometrics.

Asset and Liability Portfolio of Farmers

Micro Evidences from India

S. Bisaliah, S. Mahendra Dev,
Syed Saifullah & Dhritisree Sarkar

ACADEMIC FOUNDATION
NEW DELHI

www.academicfoundation.com

First published in 2014
by

ACADEMIC FOUNDATION
4772-73 / 23 Bharat Ram Road, (23 Ansari Road),
Darya Ganj, New Delhi - 110 002 (India).
Phones : 23245001 / 02 / 03 / 04.
Fax : +91-11-23245005.
E-mail : books@academicfoundation.com
www.academicfoundation.com

in association with the
Indira Gandhi Institute of Development Research (IGIDR), Mumbai.

Cataloging in Publication Data--DK
Courtesy: D.K. Agencies (P) Ltd. <docinfo@dkagencies.com>

Asset and liability portfolio of farmers : micro evidences from India / S. Bisaliah ... [et al.].
p. cm.
Includes bibliographical references.
ISBN 9789332701366

1. Farmers--India--Economic conditions. 2. Agriculture--India--Finance. 3. Asset-liability management--India. 4. Household surveys--India. I. Bisaliah, S.

DDC 338.10954 23

Typeset by Italics India, New Delhi.

Printed and bound by The Book Mint, New Delhi.
www.thebookmint.in

Contents

List of Tables and Figures

Tables

Figures

Preface

The book, *Asset and Liability Portfolio of Farmers: Micro Evidences from India*, is based on research conducted by the authors, using farm-level database generated by the National Sample Survey Organisation (NSSO), Government of India and National Council of Applied Economic Research (NCAER), New Delhi. Farm business asset in the form of agricultural capital stock has drawn the attention of researchers for quite sometime, and there is considerable research work as related to capital formation in/for agriculture, farm output growth and poverty alleviation. But the present study treats agricultural capital stock for farm business activities as only one type of capital accumulation by farm households, the other three types being non-farm business asset, household asset and financial asset. An empirical study of these four broad types of assets and their composition would provide insights into farmers' preference structure in asset accumulation, and thereby policy directions needed. An attempt is made in this research study to address six basic issues. First, how is the asset-liability portfolio structure of farmer households, and whether there have been compositional shifts in their portfolio structure? Second, what factors would account for likelihood of financial inclusion and exclusion of farm households? Third, how is the pattern of capital expenditure and its determinants? Fourth, what factors would govern the accumulation of non-farm business assets, and thereby non-farm business activities for livelihood options? Fifth, are there some fundamental differences between large and small farmers with respect to some of these issues? Sixth, what are the policy directions derived from the inferences drawn from the analysis of two data sets?

In conducting this study, the sponsorship of Food and Agriculture Organisation (FAO), Rome, help of Dr Madhur of World Bank, Washington, in providing NCAER Rural Economic and Demographic Survey (REDS) data for the year 2005-06 and help of Ms Kalavathi in providing secretarial assistance are gratefully acknowledged.

—S. Bisaliah, S. Mahendra Dev,
Syed Saifullah, Dhritisree Sarkar

1 The Theme, Issues Addressed and Study Objectives

Statement of the Study Theme

In aggregative terms, India's overall economic performance during the last two decades or so has been impressive. After decades of slow and low gross domestic product (GDP) growth, India has finally turned the corner to be labelled as one of the fastest growing economies in the world, like China. But unlike China, the emerging growth pattern (accelerated growth in service sector, stagnant/marginally accelerated growth of industrial sector and decelerated growth of agriculture sector) of India has led to more of 'exclusive growth' than 'inclusive growth.' Hence the 'exclusive growth' pattern as well as the need for 'inclusive growth' has given rise to the challenge of identifying sectors, regions and socioeconomic groups which are outside the growth/development orbit, and of identifying the needed policies/programmes to address the problem of enclave growth so as to make growth/development broad-based and virtuous. Agriculture and allied sector is identified as one of the major areas of focus for making growth more inclusive (Dev, 2008). Not withstanding the decline in the share of agriculture to national GDP, accelerated agriculture growth is the main pathway out of poverty for poor people of India; for addressing broader development policy goals such as food security, employment generation and bridging rural-urban development divide; and for forming a critical edifice of the strategy for overall development of the economy. This is evident from the Chinese model of agriculture/rural development, emphasising that agriculture is not an inferior partner in overall development, neglect of agriculture can slow down the growth of national economy and realisation of other development policy goals, and that growth of grain production is a necessary condition for food security and social stability.

Agriculture Growth Cycle

Indian agriculture sector has been passing through a growth cycle—peak growth rate of about 3.5 per cent in 1980s, decelerated growth rate of around 2.6 per cent from the mid-1990s to mid 2000s, marginal recovery and slow down. This kind of growth cycle questions the sustainability of Indian agriculture in the face of external and internal shocks (Madhur *et al.*, 2011), pointing out towards the inability of the sector to meet rising demand for agri-food products, to contribute for moving the national economy to high growth trajectory, to mimise rural-urban development divide, to contain rising food inflation on a sustainable basis and to meet the Millennium Development Goal (MDG) of halving poverty by 2015.

In view of predominance of agriculture sector and the 'growth tensions' which Indian agriculture has been passing through, search for identification of constraints on Indian agriculture and the constraint releasing policies and strategies has led to many problems to be addressed. First, low as well as slowing down of growth of agriculture productivity is the main problem. Since 1990s, productivity growth has weaked, after experiencing high productivity growth during the green revolution era. Needless to say that productivity growth is important to temper income disparities and social tensions. Second, some farm enterprise diversification has taken place, but diversification is insufficient with implications for food/nutrition security and income security. The present trend towards diversification is not enough to keep pace with changing consumption pattern. How much of productivity growth across the board is needed to meet the projected demand and changing consumption pattern? Third, agricultural development segmentation as reflected in regional agricultural development disparities, disparities in the productivity growth of crop groups like cereals and pulses, productivity disparities at farm level as between commercial and subsistence farms and irrigated and rainfed farms, and as between different farm-size groups is quite an issue to be addressed for a broad-based and more inclusive agricultural growth pattern. Fourth, how to address the problem of sluggish transition of labour out of agriculture into non-farm sector, and how to increase the productivity of labour in agriculture? Fifth, what are the additional reforms needed in areas such as land ownership, input-output market, credit system and subsidies? Sixth, to address the problem of productivity differential, how important is technical efficiency for some farm households (like subsistence group) and technical progress for other farm households (like commercially oriented group)? What are the search areas for increasing technical efficiency and technical progress in growth domain? Seventh, low and declining total factor productivity (TFP) of Indian agriculture is well documented (Chand *et al.*, 2011). It is also inferred that output growth in India has been dominantly input-led growth. The agriculture growth pattern required for India is one of TFP-led growth, which is quite crucial for reducing cost per unit of farm output and for making agriculture competitive. The determinants of TFP growth are also documented (Bisaliah *et al.*, 2011; Chand *et al.*, 2011), of which investment in and for agriculture is a frontline determinant. Eighth, yet some other contextual and complexities of Indian agriculture need to be placed in this perspective—small and marginal farms forming over 80 per cent of total holdings, productivity (yield) fatigue, agricultural technology fatigue, depletion and degradation of natural resource base, and low capital base (including rural human capital). Ninth, yet another issue is the relevance of cross-country business models at both macro and farm levels (Bisaliah, 2012) to accelerate capital formation in/for agriculture.

Hence, there are many search areas to place agriculture on high growth trajectory (Bisaliah, 2012). Most of these search areas need investment/capital formation in agriculture. But one of the most disquieting developments in the agriculture sector during the last two decades or so has been the neglect of capital formation in agriculture, especially in public sector (Bisaliah *et al.*, 2011; Bisaliah, 2012). The experience of China shows that high rate of public invest in agriculture can ensure big spurt in agriculture output (Bisaliah, 2012). Documentation and analysis of investment/capital formation in Indian agriculture, using both macro and micro (farm level) data, has arrived at four major inferences. First, the investment growth cycle during the last six decades in Indian agriculture encompasses a period of rising trend, a subdued phase, a peak epoch, a persistent deceleration and perceptible recovery. Second, the share of public sector in total investment had declined from 45 per cent in 1980s to about 20 per cent in 2000s, whereas the share of private sector had increased to about 80 per cent in 2000s. Out of total private investment in agriculture, farm household investments contribute over 90 per cent. Third, the macro evidences have led to the conclusion that high rate of growth of capital stock in agriculture would lead to high rate of growth in agricultural productivity and poverty reduction. Further, evidences from developing countries also draw the inference that there is a positive association between agricultural capital stock and productivity of labour engaged in agriculture and negative association between agricultural capital stock and productivity of labour engaged in agriculture and negative association between productivity of labour and prevalence of hunger and poverty (Stephan *et al.*, 2009). Fourth, the analysis of farm-level data has led to the inferences that land has formed over 90 per cent of on-farm total capital stock, implying inadequacy of non-land capital assets such as animal capital, irrigation capital and farm machinery and implements, and between 1994-95 and 2007-08, there was a negative/low growth rate of non-land capital assets as a group, and that non-land capital stock per hectare of land and per unit of labour had declined during the same period of 13 years (Bisaliah, 2012).

The preceding section is just an extraction of main inferences on macro and micro study of capital formation in Indian agriculture. Micro evidences are based on farm-level data. But farm household as an economic entity has its own investment portfolio with respect to its assets of which agriculture capital assets would form only one type of asset. In their investment portfolio, farm households have four kinds of assets, *viz.*, farm business assets (called agriculture capital stock so far), non-farm business assets, household durable assets and financial assets, the definition and measurement of which will follow in Chapter 2. The present study is an attempt to address the issues related to investment portfolio of rural and urban cultivator households empirically.

There are five issues which need to be empirically addressed, using farm household level data on investment portfolio. First, how is the investment portfolio of cultivator households at a point in time as well as over a period of time? Second, what would constitute the growth pattern of assets over a period of time? Third, what kind of linkage could be observed between composition and growth of assets and credit availment behaviour of cultivator households, with an emphasis on financial inclusion/exclusion? Related to this is the issue of variations in asset-debt ratio of cultivator households and factors which would explain this variation in debt-asset ratio? Fourth, how are capital expenditure on various assets and debt-intensity related? Fifth, what factors would govern asset transaction—

acquisition and depletion (sale, loss and discarded assets) of assets? Could there be a pattern (patterns) of asset transaction behaviour of farmers, and how these patterns are associated with credit propensity of households as well as their 'asset status' and its composition? All these issues need to be addressed across farm-size groups, across rural and urban cultivator households and across cluster of states/regions with high, moderate and low agricultural productivity status.

Specific Objectives of the Study

The issues identified for analysis are translated into specific objectives as below:

- Analyse investment portfolio structure of cultivator households and its composition in terms of farm business, non-farm business, household and financial assets as well as the components of these four broad groups of assets.
- Estimate the growth rates of different kinds of assets and identify the pattern of growth.
- Identify determinants of non-farm business assets accumulation.
- Analyse household liability structure in terms of debt portfolio, debt-asset relations, factors governing financial inclusion and factors explaining variations in debt-asset ratio.
- Examine the capital expenditure structure, growth, capital expenditure and debt relations, and identify determinants of capital expenditure.
- Derive pathways and policy directions for 'influencing' farm household investment portfolio.

As discussed earlier, all these objectives of the study are subjected to available empirical evidences on cultivator households across farm-size groups, rural and urban households, and cultivator households of three 'clusters' of states based on agricultural productivity criteria. Chapters 3, 4 and 5 analyse asset and liability structure and capital expenditure pattern, based on the National Sample Survey Organisation (NSSO) data for the years 1991-92 and 2002-03. Based on data collected by the National Council of Applied Economic Research (NCAER) for the years 1998-99 and 2005-06, the analysis is performed and results are discussed in Chapter 6.

2 Data Sources, Sampling Methodology, and Definition and Measurement of Concepts

Given the objectives set out for the study in Chapter 1, the present chapter is concerned with discussion on data sources, sampling methodology followed in data generation through survey and definition and measurement of key concepts.

Data Sources: Historical Perspective

Historically, the first attempt was made by the Reserve Bank of India (RBI) to study both the demand and supply sides of credit in its report "All India Rural Credit Survey: 1951-52". This survey focused on demand for credit by farm families and supply of credit by both institutional and non-institutional agencies. Similarly, during 1961-62, the RBI conducted "All India Debt and Investment Survey" (AIDIS). The scope of this survey was extended to include capital expenditure incurred by rural households.

The third comprehensive survey (1971-72) in this series of decennial surveys was entrusted by the RBI to the National Sample Survey Organisation (NSSO). Since 1971-72, NSSO has been conducting decennial AIDIS (NSSO, 1998, for details on different blocks of survey). This third survey integrated AIDIS with land and livestock holding survey and extended to cover urban households. The fourth survey for 1981-82, fifth survey for 1991-92 (NSS 48th round) and sixth survey for 2001-02 (NSS 59th round) have been conducted by NSSO. While conducting these surveys, NSSO has made some marginal changes in sample design and number of visits for data collection. But there has been no change in basic structure of data collected. These surveys have generated basic quantitative information on assets, liabilities and capital expenditure in the household sector of the economy.

Based on this data generated under AIDIS, the NSSO prepares five reports. The first report provides estimates of assets and liabilities of rural and urban households. The second report deals with several aspects of indebtedness such as intensity of cash loans, duration, rate of interest and credit agencies. The third report focuses on different aspects of flow variables of households such as current borrowing, repayment, type of loans and security. The focus of the fourth report is on selected aspects of household assets and liabilities of different social groups such as scheduled caste (SC), scheduled tribe (ST), other backward castes and religions groups. The fifth report deals with broad features of capital expenditure and asset transaction, *viz.*, acquisition and depletion of assets by the rural and urban households.

Given the objectives outlined for the study, unit level (household level) data on various kinds of assets and liabilities (mainly credit status), asset transaction in a particular frame of time (say one year) and capital expenditure incurred by study cultivator households in a particular frame of time (say one year) are needed. Hence the study uses primary data of cultivator households collected by NSSO for the years 1991-92 and 2001-02, and primary data collected by the National Council of Applied Research (NCAER) for the years 1998-99 and 2005-06.

Sampling Methodology and Sample Structure

With respect to sample design, the NSSO has employed two-stage sampling method, with the first stage unit as village/urban blocks and the second stage unit as households in both rural and urban areas of the country. The survey method has been used for data collection from sample households through one/two visits depending on the type of data to be collected.

The survey classifies rural households broadly into cultivators and non-cultivators (agriculture labour and artisans like weavers, tailors carpenters and so on), and urban households into self-employed, regular wage/salaried employees and others. Further, the urban self-employed households are engaged in the farm or non-farm enterprises. For the present study, all the cultivator households of both rural and urban areas are selected from the total pool of households surveyed.

Further, the cultivator households (both rural and urban) are classified into three farm groups—marginal and small operating upto 2 ha of land (S), semi-medium and medium operating upto 10 ha (M) and large operating more than 10 ha (L).

Yet, the present study has developed another sub-sample, *viz.*, based on agriculture productivity. Based on agriculture productivity (a three-year average) criterion, the states are divided into three clusters with two states in each cluster—Punjab and Tamil Nadu as high-productivity cluster, Uttar Pradesh and Bihar as mid-productivity cluster and Madhya Pradesh and Rajasthan as low-productivity cluster.

The sample structure of the present study based on NSSO data is given in Table 2.1. The total sample comes to 24,035 for the year 1991-92 and to 48,295 for the year 2001-02. The sampling methodology, and sample structure and size of NCAER data will be discussed in Chapter 6.

Table 2.1

Sample Structure of Cultivator Households

Sample	*1991-92*				*2002-03*			
	Small	*Medium*	*Large*	*Total*	*Small*	*Medium*	*Large*	*Total*
India	17426 (72.50)	6135 (25.53)	474 (1.97)	24035 (100)	41208 (85.33)	6718 (13.91)	369 (0.76)	48295 (100)
Rural	15511 (71.48)	5755 (26.52)	434 (2.00)	21700 (100)	38597 (85.67)	6151 (13.65)	306 (0.68)	45054 (100)
Urban	1915 (82.01)	380 (16.27)	40 (1.71)	2335 (100)	2611 (80.56)	567 (17.49)	63 (1.94)	3241 (100)
LPR								
Rural	1614 (49.66)	1433 (44.09)	203 (6.25)	3250 (100)	5120 (70.87)	1959 (27.11)	146 (2.02)	7225 (100)
Urban	172 (63.24)	87 (31.99)	13 (4.78)	272 (100)	348 (65.05)	167 (31.21)	20 (3.74)	535 (100)
MPR								
Rural	5037 (81.77)	1103 (17.91)	20 (0.32)	6160 (100)	13271 (93.19)	951 (6.68)	19 (0.13)	14241 (100)
Urban	381 (85.23)	63 (14.09)	3 (0.67)	447 (100)	709 (91.37)	65 (8.38)	2 (0.26)	776 (100)
HPR								
Rural	1172 (68.30)	510 (29.72)	34 (1.98)	1716 (100)	2761 (85.19)	451 (13.92)	29 (0.89)	3241 (100)
Urban	234 (82.69)	46 (16.25)	3 (1.06)	283 (100)	344 (81.13)	73 (17.22)	7 (1.65)	424 (100)

Note: Figures in the parenthesis are the percentages to the respective total of farm groups.
LPR: low-productivity region; MPR: medium-productivity region; HPR: High-productivity region.

Source: Estimated by authors.

Concepts and Measurement

This section focuses on the major concepts used in this study and their measurement as per NSSO methodology, and Chapter 6 will discuss these issues based on NCAER methodology.

- Cultivator household: All households operating at least 0.002 ha of land during the last 365 days preceding the date of survey are considered as cultivator households.
- Assets: One of the major objectives of AIDIS is to generate estimates of assets and liabilities; hence the need for proper classification, definition and measurement of assets, followed by the discussion on liabilities. The first report of NSSO with AIDIS data deals with the estimates of assets and liabilities of rural and urban households, as discussed under the section, Data Sources: Historical Perspective.
  - Assets (measured in ₹) are broadly categorised into physical, financial and durable household assets.
  - The physical assets are classified into farm business assets (capital stock for farm production and primary processing activities) and non-farm business assets (capital stock for non-farm enterprises).
  - Farm business refers to cultivation and processing of farm produce on the farm like paddy and gur making, and activities like livestock rearing, fishery, dairying, beekeeping and so on which are ancillary activities to agriculture.
  - Non-farm business refers to all household economic activities other than those covered in farm business. This covers manufacturing, mining and quarrying, trade, hotel, transport, construction, repairing and other services.
  - Household durable assets (measured in ₹) are those which are used for domestic purposes and have a longer life of one year and more, excepting minor assets like knives and bottles.
  - Financial assets: Financial assets (measured in ₹) are the assets like shares and deposits, dues receivable on loans advanced in

cash or kind, deposits in companies, banks, post offices and cash on hand.

- Valuation of assets: Physical assets are valued at the current market price of the asset in the existing condition, prevailing in the locality. In case of disposed assets, the sale price has been considered.

  The market values of physical assets are deflated by using the deflators developed by Central Statistical Organisation (CSO) with 2004-05 as the base year so as to facilitate the estimation of growth rates of physical assets over a period of time.

- Liabilities: Household liabilities (measured in ₹) include loans (cash and kind, and unpaid bills) and outstanding taxes.

  Loans are classified into institutional and non-institutional. The purpose of loan includes capital and current expenditure in farm business and non-farm business, and financial investment expenditure.

  Further, the loans are divided into short-term loans (upto 12 months), medium-term loans (1-3 years), and long-term loans (more than 3 years). In this study, only medium- and long- term loans are considered, in view of their direct bearing on investment portfolio of cultivator households.

- Capital expenditure: Data relating to capital expenditure on farm business non-farm business, and some household items are available in AIDIS. It is recalled from Chapter 1 (Section: Specific Objectives of the Study) that one of the objectives of the present study is to analyse the structure of capital expenditure incurred by cultivator households. Given this, the concepts of capital expenditure and their measurement (all items of capital expenditure measured in ₹) need to be made explicit.

  - Capital expenditure in farm business refers to expenditure on purchase of land, land improvement, reclamation of land, orchards and plantations, wells and other irrigation sources, agricultural implements and machinery, transport equipment, farm houses/barron/animal sheds, purchase of livestock and other miscellaneous capital expenditure as related to farm business.

- Capital expenditure in non-farm business includes expenditure incurred on account of purchase of land and its improvement pertaining to non-farm business, construction of buildings, additions and alterations/repairs to buildings, machinery and equipment, furniture and fixture and so on.
- Capital expenditure on household assets refers mainly to purchase of residential plots, purchase/construction/additions/ alterations of building for residential purpose, purchase of durable household assets, expenditure for medical treatment/ education and so on.

It needs to be made explicit that capital expenditure on farms business assets is the one which has a direct bearing on agriculture capital stock in case of cultivator households.

3 Asset Portfolio Analysis

Structure, Compositional Shifts and Growth

It is recalled from Chapter 1 that the first two objectives of the study are to analyse investment portfolio of cultivator households, shifts in composition of assets over a period of time, and to estimate the growth rates of different kinds of assets as well as to identify growth patters. Given these objectives, three basic issues have been addressed in this chapter. First, how is the structure of cultivator household asset portfolio both in aggregative and disaggregative frame, and what kind of compositional shifts have taken place over a period of time between 1991-92 and 2002-03? Second, what are the pace and patterns of asset growth over a period of time? Third, what are the determinants of asset accumulation? These three basic issues are addressed with empirical evidences from total sample, from sub-samples of rural and urban cultivators and from sub-samples of three regions based on agricultural productivity, *viz.*, low-productivity region (LPG), medium productivity region (MPG) and high productivity region (HPG). Further, in addition to overall results, empirical evidences across three farm groups, *viz.*, small and marginal (S), semi-medium and medium (M) and large farms (L) have been analysed. Further temporal variations in compositions and growth of assets owned by cultivator and non-cultivator households in rural India are also analysed.

Cross-Country Household Asset Portfolio Choices: Concepts and Evidences

The objectives set out for this chapter and the basic issues outlined for examination are placed in the perspective of concepts used and evidences generated in some cross-country studies.

- Identifying what the poor have, rather than what they do not have, focuses on their assets. The asset portfolio composition includes financial and non-financial assets. The poor are the managers of complex asset portfolio (Moser, 2006). Further, the lack of household asset is treated both a symptom and cause of poverty.
- Asset ownership and vulnerability are related. In fact, development is a process through which people's physical/material, social/organisational and motivational/attitudinal vulnerabilities (or capacities) are reduced or increased. It could be postulated that the more assets people have, the less vulnerable they are, and the greater the erosion of assets, the greater is their insecurity (Moser, 2006).
- Asset ownership is an important part of any study of social inequality. In general, the distribution of assets (both tangible and intangible) is the key determinant of the income distribution. Unequal distribution of assets also affects the equal distribution of opportunities for building both physical and human capital assets in the future. Following this, household assets are a proxy for social and economic well-being of a household. In fact, asset-based social policy is a new cutting edge development agenda, and household asset-building is a better substitute for depending heavily on welfarist approach to household income management (Moser, 2005). The concept of asset-based livelihood has led to the development of framework for sustainable livelihood approach (Ellis, 2000).
- Asset-income poverty relationships would provide empirical basis for policy design to 'influence' pattern of growth in asset accumulation (Haveman and Wolff, 2004). Further, asset poverty is a measure of economic hardship, and it is complementary to the more commonly used concept of income poverty (see Haven *et al.*, 2004, for empirical evidences on asset and income poverty over a period of time in the US). In this policy design to 'influence' the pattern of asset portfolio and their growth, asset attributes such as productivity, utility, security, holding costs, expected life, convertability and complementarity are crucial. Further these attributes would influence collateral for borrowing, for meeting

current consumption needs and increasing future income. In fact, asset portfolio choices have important implications for the rate at which wealth (networth, i.e., the difference between value of assets and the value of liabilities) is accumulated (Deborah *et al.*, 2008). Hence determinants of asset composition and networth assume importance.

- Naschold (2009), based on panel data generated under village level studies by ICRISAT (International Crops Research Institute for the Semi-Arid Tropies, Hyderabad, India), has conducted a study on structural changes in asset holding by households. Among other reasons, two major reasons are given why a study of household asset holding is relevant. First, the economic well-being of a household is dependent on its stock of assets and its composition. If dynamic perspective is kept in view, enough income to move out of poverty is possible only through accumulation of asset, and the policy of social protection may be an inevitable and should be the last option. Thus, an asset-based household welfare is more suitable for forward-looking policy design. Second, asset levels fluctuate less from day-to-day than income. Asset can be interpreted as a measure of the underlying structural well-being of a household, whereas income and to a lesser extent consumption contain a much larger amount of stochastic variation (Carter *et al.*, 2001). Using different classes of assets into a single household composite asset index, one has to use livelihood regression (Adato *et al.*, 2006). This livelihood regression expresses household well-being as a function of household characteristics and asset holdings.

Two results from Naschold (2009) study on Indian villages would provide empirical insights into dynamics of household asset holding and poverty. First, time series panel data analysis of total samples suggests economic statis in study villages. Hence these rural Indian villages cannot hope to escape poverty and move towards higher equilibrium level of asset holdings. Their survival will have to be ensured through social safety nets, the last option as discussed earlier. Second, the disaggregative analysis of same data into sub-groups leads to a patterns of what Naschold (2009) calls club convergence. Higher castes, larger holdings and more educated households are in a significantly greater asset equilibrium than their lower caste, smaller land holder and less educated peers (more on some of these patterns are given later in this chapter).

- With respect to asset portfolio management, there is an analogy between households and corporate firms, and there are differences also. Asset accumulation would also involve liabilities of households, i.e., funds mobilised from creditors. Hence the three items in the balance sheet of households are: assets, liabilities and wealth.
- Given some degree of substitution between different types of assets, there is always a portfolio interaction between real and financial assets, implying the nexus between physical and financial assets with access to/availment of credit (Michael and Michaelides, 2000).

In this perspective, an identification of determinants of portfolio choice such as age, wealth and education holds the key for policy, institutional and technological interventions to 'influence' the composition of assets as well as their growth.

- The farm households have multiple forms of assets and liabilities. Among these assets, farm business assets form a major/dominant component, and agricultural land is usually the main component among farm business assets. It is passed on to next generation, with important consequence for the distribution of assets in the society, which could also be determined by land policy.
- In case of households, it is not merely total asset holdings which matter but also choice of asset-mix (Chandra, 2004). Households tend to follow the famous dictum: "Don't put all your eggs in one basket." In the absence of adequate institutional mechanisms for coping with risks, households adopt asset diversification (Korir, 2011). Asset diversification by farm households is to stabilise income flows through income diversification. In fact asset, activity and income diversification lie at the heart of livelihood strategies in rural Africa (Barrett *et al.*, n.d.). Households are guided by multiple motives in their asset diversification, activity diversification and income diversification. There are two sets of motives which govern the asset-activity-income diversification (Barrett *et al.*, n.d.). First,

risk reduction and response to diminishing factor returns in any given use are termed as 'push factors.' Second, realisation of complementarity between activities such as crop-live-stock mix is termed as 'pull factor.' Added to these are the three broad groups of factors which would govern structure of household portfolio and its diversification (Gollier, 2000): risk aversion and human capital, objective of asset accumulation, and environmental such as return on investment and access to credit market.

Yet another way to explore rationale for asset diversification is to treat asset diversification as a form of self-insurance to reduce income and consumption risks. Further, the concept of economies of scope would provide another rationale for asset diversification. Economies of scope exist when the same inputs generate greater per unit profits when spread across multiple outputs than 'dedicated' to any one output (Barrett *et al.*, n.d.). Economies of scale tend to favour specialisation, whereas economies of scope favour asset and activity diversification and thereby income diversification. Hence assets, activities and income are the complementary measures in the study of diversification behaviour of households. But entry barriers (like cooked food prepared by some social groups in India is not bought by some social groups) tend to leave the poor with less diversified asset and income portfolio, leading to asset trap.

- It is recalled from Chapter 2 that NSSO provides data on four broad groups of assets, *viz.*, farm business, non-farm business, household and financial assets. Non-farm investments like accumulating and using non-farm business assets is one of the diversification strategies adopted by households. But little is known about non-farm business assets held by farmers and the variables influencing the accumulation of these assets (Serra *et al.*, 2004). There are many reasons for which non-farm business assets as well as non-farm business activities offer a ladder out of poverty. First, non-farm investments like flour mill, transport, sale of cooked food, carpentry, bicycle repair and even livestock rearing could minimise the household income variations. Further, the study by Korir (2011) has demonstrated the risk reduction benefits of non-farm business investments. Second, there is a positive relationship between non-farm income and household welfare. Rural non-farm economy enjoys faster income growth, thereby providing resources to plough back into expanded non-farm activity. This provides ladder for those with skills, some amount of financial resources, education and social contacts to take advantages of emerging market opportunities. Entry to high return niches like non-farm business activities in rural areas is possible for the households endowed with these resources and skills. Third, non-farm activity is becoming more important in rural Africa, contribution 40-45 per cent of average household income (Barrett *et al.*, n.d.). Hence there are some empirical regularities to indicatc positive relationship between non-farm investment as well as income and household welfare across most of the rural Africa.

There are many determinants of non-farm investment/asset such as age, education, net worth, credit accessibility/availment and family size of household. For example, in accordance with the life cycle theory, farm operators age is a relevant variable when trying to explain non-farm business investment and non-farm business asset accumulation (Serra *et al.*, 2004).

Aggregative and Disaggregative Analysis of Assets Structure

The focus of this section is to analyse the results on structure of farm household assets with results derived from:

- All-India data set.
- Rural and urban sub-samples.
- Data set pertaining to productivity-based three agricultural regions in the country.
- Sub-data-set of three farm groups across all-India sample, rural and urban sample, and regional sample.
- Further, the intertemporal comparisons of asset structure as between 1991-92 (Period I; P.I) and of 2002-03 (Period II; P.II) with respect to total and sub-data-sets.

All-India Aggregative and Disaggregative Results

Table 3.1 provides results on percentage share of four broad groups of assets, *viz.*, farm business (FBA), non-farm business (NFBA), household (HHA) and financial assets (FA) during the two periods in total assets owned by cultivator households across the country. First, the dominant asset group in overall is FBA, followed by HHA. The percentage share in total assets is very low in case of NFBA and FA. Second, the percentage share of FBA varies directly with farm size, i.e., highest share in case of large farms (as high as 79%) and lowest share (as low as 52%) in case of small farms. Third, in case of NFBA, the highest share (6%) is found in case of small farms, and a share of 4 per cent each in case of other two farm groups. Fourth, the percentage share of HHA in total assets varies inversely with farm size, the share being the highest (41%) in case of small farms. Fifth, in case of all

Table 3.1

Composition of Assets and Temporal Shifts: All India

Percentage Share

Assets	*S*		*M*		*L*		*Overall*	
	II	*I*	*II*	*I*	*II*	*I*	*II*	*I*
FBA	52 (-)	53	73 (-)	75	79	79	60 (-)	66
NFBA	6 (+)	4	4 (+)	3	4 (+)	3	5 (+)	4
HHA	41	42	23	22	17	18	34	31
FA	1 (+)	0	0	0	1 (+)	0	1 (+)	0
Total Assets	100	100	100	100	100	100	100	100

Note: FBA = farm business assets;
NFBA = non-farm business assets;
HHA = household assets;
FA = financial assets;
I = 1991-92, II = 2002-03;
S = marginal and small farms (<2 ha);
M = semi-medium and medium farms (>= 2 and <10 ha);
L = large farms (>= 10 ha)
(+) = Percentage share has increased in 2002-03 compared to 1991-92
(-) = Percentage share has decreased in 2002-03 compared to 1991-92 and no sign represents no changes.

Source: Estimated by authors.

farm groups, the share of FA is insignificantly low. Sixth, with respect to inter temporal shifts in composition of assets between two periods of time, the overall results suggest some gain in percentage share of NFBA, HHA and FA, and some loss in FBA. With respect to farm groups, all the groups stand to gain in case of NFBA. This result is suggestive of the inference that cultivator households have made a development move, however small the move is, towards accumulation of NFB category of assets. This is a symptom of transition towards non-farm activity to reduce variance in their income and consumption patterns. Further, in case of assets in the category of FA, there is a symptom of gain during P.II compared to P.I in case of all the three farm groups.

The next step is to analyse all-India disaggregative results on all four groups of assets (Tables 3.2-3.4).

With NSSO data, five components of FBA are used for disaggregative analysis. It could be seen from Table 3.2 that:

- Out of total FBA, the overwhelmingly dominant share is that of land (either in over all terms or in farm group-wise terms), contributing around 92 per cent. Added to this, the share of land in total FBA has increased during P.II compared to P.I. This implies that the share of non-land assets is very low. This reinforces the result derived from the study (Bisaliah and Dev, 2010), using data generated under Comprehensive Cost of Cultivation Scheme of Government of India. Among non-land capital assets, the share of agricultural machinery and irrigation capital is found to have decreased during P.II in case of both M and L categories of farms. Fall in the share of animal capital in case of small farms could be considered as a disturbing trend, because income from animal enterprises is expected to be supplementary and complementary to income from crop enterprises of these farms. However, it is likely that animal power is replaced by mechanical power (perhaps through custom hiring) for agricultural operations. In case of farm business buildings and transport equipment, not much of a change is observed as between the two periods of time.

Table 3.2

Disaggregative Analysis of Portfolio of Farm Business Assets: All India

Percentage Share

Components	*S*		*M*		*L*		*Overall*	
	II	*I*	*II*	*I*	*II*	*I*	*II*	*I*
L	93 (+)	92	92 (+)	91	93 (+)	90	93 (+)	92
B	3	3	2	2	1 (-)	2	2	2
AM and IC	3	3	4 (-)	5	5 (-)	6	4	4
AC	1 (-)	2	1	1	0 (-)	1	1	1
TE	0	0	1	1	1	1	1	1
Total	100	100	100	100	100	100	100	100

Note: FBA = farm business assets;
L = land of farm business;
B = building of farm business;
AM and IC = agricultural machinery and irrigation capital;
AC = animal capital for farm business;
TE = transport equipment for farm business.

Source: Estimated by authors.

It could be observed from Table 3.3 that:

- Among non-farm business assets, the most dominant one is animal capital for non-farm business, and the share is about two-third in case of S and M category of farms. Added to this is the increased share of animal capital (68%) in total NFBA owned by small farms, perhaps indicating the increased importance of animal capital (and thereby animal husbandry) under NFBA, whereas it is found to be very low under FBA category (Table 3.2).
- The share of building for non-farm business, and of machinery and transport, even though important in total share in NFBA, has experienced a fall in their share. The fall in their share is much more in case of S farms. In case of L farms, there has been a fall in the share of building for non-farm business and of non-farm business machinery during the second period.

Table 3.3

Disaggregative Analysis of Portfolio of Non-Farm Business Assets: All India

Percentage Share

Components	*S*		*M*		*L*		*Overall*	
	II	*I*	*II*	*I*	*II*	*I*	*II*	*I*
L	3	NA	8	NA	7	NA	4	NA
B	13 (-)	25	15 (+)	12	17 (-)	21	14 (-)	20
AC	68 (+)	44	62	62	46 (+)	44	66 (+)	50
NBM	7 (-)	18	8 (-)	15	8 (-)	16	7 (-)	17
TE	9 (-)	12	8 (-)	12	21 (+)	19	9 (-)	12
Total	100	100	100	100	100	100	100	100

Note: NFBA = non-farm business assets;
L = land for non-farm business;
B = building for non-farm business;
AC = animal capital for non-farm business;
NBM = non-farm business machinery;
TE = transport equipment for non-farm business.

Source: Estimated by authors.

Table 3.4 displays results on the percentage share of four components of HHA and shifts in their share during the study periods.

- In the total HHA, building (residential) claims the highest share (53%), followed by land (33%), for household activities (other than the land under FBA), household durable goods (12%), and transport equipment (2%) for household use. The farm groups follow the same pattern of percentage share of four components.
- Compared to the first period, there has been a gain in the share of land for household during the second period in case of all three farm groups, but loss in the share of buildings for household as well as household durable goods. Due to change in 'life style,' the share of transport equipment for household use has either remained constant (as it is in case of L farms households) or increased (as in the case of S and M categories of farm households).

The percentage share of different kinds of financial assets is not analysed in view of the insignificant share of financial assets in total assets of cultivator households, as shown in Table 3.1.

Table 3.4

Disaggregative Analysis of Portfolio of Household Assets: All India

Percentage Share

Components	*S*		*M*		*L*		*Overall*	
	II	*I*	*II*	*I*	*II*	*I*	*II*	*I*
L	34 (+)	28	31 (+)	25	35 (+)	32	33 (+)	27
B	53 (-)	54	53 (-)	55	46 (-)	48	53 (-)	54
TE	2 (+)	1	3 (+)	2	5	5	2	2
DG	12 (-)	17	13 (-)	18	14 (-)	15	12 (-)	18
Total	100	100	100	100	100	100	100	100

Note: HHA = household assets;
L = land for household;
B = building for household;
TE = transport equipment for household;
DG = durable goods.

Source: Estimated by authors.

Rural Cultivator Households: Aggregative and Disaggregative Results

The results derived from the all-India total sample have been analysed in the section preceded. In the present section, two sub-samples, *viz.*, rural (this section) and urban cultivator households (the next section), sorted out from the all-India total sample, are subject to empirical analysis of asset structure in both aggregative and disaggregative frame.

In aggregative terms, the structure and compositional shifts of assets owned by rural cultivator households are analysed in Table 3.5.

- In overall terms, during both the periods, the highest share (70%-63%) to total assets of four groups is by FBA, followed by HHA

(32%-27%). The share of NFBA varies between 5 per cent to 3 per cent, and the share of financial assets is as good as zero. There has been an increase in the share of NFBA as well as HHA during P.II compared to that of P.I, whereas FBA has experienced a decline in its share.

- With respect to farm-size wise results, both S and M groups of farms stand to loose in the share of FBA, but they stand to gain in the share of NFBA. It is a possible 'hunch' to think that both S and M groups of rural households have started building up their NFBA to diversify their asset portfolio 'in favour' of non-farm business activities and thereby diversifying their income sources.

Table 3.5

Composition of Assets and Temporal Shifts: Rural Households

Percentage Share

Assets	*S*		*M*		*L*		*Overall*	
	II	*I*	*II*	*I*	*II*	*I*	*II*	*I*
FBA	55 (-)	60	75 (-)	77	84 (+)	83	63 (-)	70
NFBA	6 (+)	3	4 (+)	2	3 (+)	3	5 (+)	3
HHA	39 (+)	37	21	21	13 (-)	15	32 (+)	27
FA	1 (+)	0	0	0	0	0	0	0
Total Assets	100	100	100	100	100	100	100	100

Note: Definitions of abbreviations are as under Table 3.1.

Source: Estimated by authors.

The disaggregative results on all the components of FBA, NFBA, HHA and FA owned by rural cultivator households form the theme of this section.

It could be observed from Table 3.6 that land alone contributes over 90 per cent share to total FBA group. The next highest share of 4 per cent is in the form of agricultural machinery and irrigation capital, followed by buildings for farm business, animal capital and transport equipment.

- All the three farm groups are found to follow the pattern of highest share (over 90%) contributed by land, followed by AM and IC, B, AC and TE category of assets.

- Over a period of time excepting the increased share of land (marginally) in overall as well as in farm groups, no substantial difference in the share of component of FBA is observed.

Table 3.6

Disaggregative Analysis of Portfolio of Farm Business Assets: Rural Households

Percentage Share

Components	*S*		*M*		*L*		*Overall*	
	II	*I*	*II*	*I*	*II*	*I*	*II*	*I*
L	93 (+)	92	92 (+)	91	93 (+)	90	92 (+)	91
B	3	3	2	2	1 (-)	2	2	2
AM and IC	3	3	5	5	5 (-)	6	4	4
AC	1 (-)	2	1	1	0 (-)	1	1	1
TE	0	0	1	1	1	1	1	1
Total	100	100	100	100	100	100	100	100

Note: Definitions of abbreviations are as under Table 3.2.

Source: Estimated by authors.

In case of components of NFBA group (Table 3.7), animal capital asset alone contributes 74 per cent during the second period, with 12 per cent to 6 per cent contribution by building for non-farm business, transport equipment and non-farm business machinery.

- In case of M and L group of farms, there has been a gain in the share of building for non-farm business, during the second period, but loss in case of S group. In case of animal capital components, S and L group of farms stand to gain. But both S and M group of farms stand to loose in the share of NBM as well as TE.

A disaggregative analysis of results (Table 3.8) on portfolio of HHA suggests three major inferences:

- In overall, during both the periods, the share of residential building in total household assets is the dominant (57%-55%) one, followed by land (30%-24%) and household durable goods (18%-13%), with TE forming a minor component.

- The same pattern of 'dominant asset' or 'otherwise' is observed in case of all farm groups.

Table 3.7

Disaggregative Analysis of Portfolio of Non-Farm Business Assets: Rural Households

Percentage Share

Components	*S*		*M*		*L*		*Overall*	
	II	*I*	*II*	*I*	*II*	*I*	*II*	*I*
L	1	NA	1	NA	0	NA	1	NA
B	12 (-)	16	13 (+)	10	17 (+)	11	12 (-)	13
AC	75 (+)	69	72 (-)	73	68 (+)	64	74 (+)	70
NBM	5 (-)	8	7 (-)	9	6	6	6 (-)	8
TE	7 (-)	8	7 (-)	8	9 (-)	18	7 (-)	9
Total	100	100	100	100	100	100	100	100

Note: Definitions of abbreviations are as under Table 3.3.
Source: Estimated by authors.

Table 3.8

Disaggregative Analysis of Portfolio of Household Assets: Rural Households

Percentage Share

Components	*S*		*M*		*L*		*Overall*	
	II	*I*	*II*	*I*	*II*	*I*	*II*	*I*
L	31 (+)	25	28 (+)	24	24 (+)	18	30 (+)	24
B	55 (-)	57	55 (-)	56	52 (-)	55	55 (-)	57
TE	1	1	3 (+)	2	6 (+)	3	2 (+)	1
DG	12 (-)	17	14 (-)	18	18 (-)	24	13 (-)	18
Total	100	100	100	100	100	100	100	100

Note: Definitions of abbreviations are as under Table 3.4.
Source: Estimated by authors.

- Over a period of time, the share of both residential building and household durable goods in total household assets is found to have decreased, whereas the share of land as well as TE is found to have increased.

As previously mentioned, the share of different components of financial assets is not analysed.

Urban Cultivator Households

The aggregative as well as disaggregative results with the analysis of asset structure and shifts in composition over a period of time in case of urban cultivator households are shown in Tables 3.9-3.12.

In overall, the highest share of total assets of urban cultivator households is in the form of HHA, followed by FBA and NFBA. Across farm groups, the share of FBA varies directly with farm size and of HHA varies inversely with farm size. In case of NFBA, the variations across farm groups is limited. Over a period of time, FBA has experienced gain in the share under S and L category of farms and fall in the share under M category. HHA has gained in its share under M category and experienced fall in its share in case of S and L category. The same experience of fall in the share of NFBA under S and L category of farms is observed. With respect to FA, there has been some gain in case of all groups of unban cultivator households (Table 3.9).

Table 3.9

Composition of Assets and Temporal Shifts: Urban Households

Percentage Share

Assets	*S*		*M*		*L*		*Overall*	
	II	*I*	*II*	*I*	*II*	*I*	*II*	*I*
FBA	34 (+)	30	58 (-)	60	64 (+)	56	45 (+)	41
NFBA	7 (-)	9	6	6	7 (-)	8	7 (-)	8
HHA	57 (-)	61	35 (+)	34	28 (-)	36	47 (-)	51
FA	1 (+)	0	1 (+)	0	1 (+)	0	1 (+)	0
Total Assets	100	100	100	100	100	100	100	100

Note: Definitions of abbreviations are as under Table 3.1.

Source: Estimated by authors.

The component analysis of FBA (Table 3.10) of urban cultivator households leads to the following inferences:

- About 94 per cent is in the form of land asset, 3-4 per cent is in the form of AM and IC group and 2 per cent in the form of buildings for farm business. The share of AC as well as TE is very limited. Between 1991-92 and 2002-03, there has been a gain in the share of land assets, and fall in the share of AM and IC and AC groups of assets.

Table 3.10

Disaggregative Analysis of Portfolio of Farm Business Assets: Urban Households

Percentage Share

Components	*S*		*M*		*L*		*Overall*	
	II	*I*	*II*	*I*	*II*	*I*	*II*	*I*
L	94 (-)	95	95 (+)	92	94 (+)	90	94 (+)	93
B	3 (+)	2	1 (-)	2	1 (-)	2	2	2
AM and IC	2	2	3 (-)	5	4 (-)	6	3 (-)	4
AC	0 (-)	1	0 (-)	1	0 (-)	1	0 (-)	1
TE	0	0	1	1	1	1	1	1
Total	100	100	100	100	100	100	100	100

Note: Definitions of abbreviations are as under Table 3.2.
Source: Estimated by authors.

In case of component analysis (Table 3.11) of NFBA:

- The share of AC (29%) dominates followed by the share of B group of assets (22%), TE (18%) and NBM (13%).
- The percentage share of AC in total NFBA varies inversely with farm size and also of NBM. In case of TE, the dominant share (39%) is found in case of large farms, followed by S and M groups of farms. The percentage share of B varies almost directly with farm size.
- Over a period of study period, there has been a gain in the share of animal capital asset and fall in the share of NBM of all farm groups. Both S and L farm categories have experienced a fall in the share of B asset, and gain in the share of TE group of assets.

In case of HHA (Table 3.12), the dominant share is in the form of L group of asset, followed by B-asset category and the category of household durable assets, with the lowest share of TE.

The land asset has increased its share over a period of time under all the three category of farms, whereas B-asset category and DG-asset category have experienced a fall in their shares.

Table 3.11

Disaggregative Analysis of Portfolio of Non-Farm Business Assets: Urban Households

Percentage Share

Components	*S*		*M*		*L*		*Overall*	
	II	*I*	*II*	*I*	*II*	*I*	*II*	*I*
L	10	NA	35	NA	18	NA	19	NA
B	21 (-)	39	23 (+)	19	17 (-)	3	22 (-)	35
AC	36 (+)	10	21 (+)	16	14 (+)	6	29 (+)	11
NBM	14 (-)	34	12 (+)	38	12 (-)	35	13 (-)	35
TE	18 (+)	17	10 (-)	28	39 (+)	21	18 (-)	20
Total	100	100	100	100	100	100	100	100

Note: Definitions of abbreviations are as under Table 3.3.
Source: Estimated by authors.

Table 3.12

Disaggregative Analysis of Portfolio of Household Assets: Urban Households

Percentage Share

Components	*S*		*M*		*L*		*Overall*	
	II	*I*	*II*	*I*	*II*	*I*	*II*	*I*
L	45 (+)	36	41 (+)	29	44 (+)	34	44 (+)	34
B	43 (-)	46	45 (-)	49	42 (-)	46	43 (-)	47
TE	2	2	4 (+)	3	4	4	2	2
DG	10 (-)	16	10 (-)	19	11 (-)	17	10 (-)	17
Total	100	100	100	100	100	100	100	100

Note: Definitions of abbreviations are as under Table 3.4.
Source: Estimated by authors.

Asset Structure and Compositional Shifts in Regions

The aggregative and disaggregative results on asset structure and their compositional shifts across three farm groups of three regions identified based on agriculture productivity are shown in Tables 3.13, 3.14 and 3.15. The major inferences from these results are the following:

- With respect to percentage share of the four broad groups of assets in total assets (Table 3.13):
  - In all three regions, *viz.*, LPR, MPR and HPR, 64-65 per cent of total assets is contributed by FAB, 29-31 per cent by HHA, and 5-6 per cent by NFBA, and the share of FA is insiginant.
  - The share of NFBA and HHA varies inversely with farm size, and of FBA directly with farm size.
  - Over a period of time, the share of FBA in total assets is found to have decreased and of NFBA as well as HHA increased in all the three regions.
  - In all the three regions, the 'dominant' asset groups are FBA and HHA.
- Component analysis of portfolio of FBA (Table 3.14) suggests that:
  - Over 90 per cent of FBA is in the form of land, 5-4 per cent in the form of AM and IC, 2-3 per cent in the form of B assets and TE forming hardly 1 per cent. Across the farm groups, almost the same pattern is observed.
  - Over a period of time, the cultivator households of both HPR and LPR are found to have experienced a fall in the share of AM and IC and an increase in the share of land in total FBA.
- With respect to components of NFBA (Table 3.15):
  - AC asset dominates with a share ranging between 77 per cent (MPR) and 53 per cent (HPR). The next dominant one is B asset with a share ranging between 13 per cent (LPR) and 10 per cent (HPR).
  - In LPR, the highest share is contributed by AC (70%) followed by B asset (13%), NBM (7%) and TE (6%).

Table 3.13

Composition of Assets and Temporal Shifts: Three Agricultural Productivity Regions

Percentage Share

Assets	LPR								MPR								HPR							
	S		M		L		Overall		S		M		L		Overall		S		M		L		Overall	
	II	I	II	I	II	I	II	I	II	I	II	I	II	I	II	I	II	I	II	I	II	I	II	I
FBA	54	54	71 (-)	74	79	79	65 (-)	70	60 (-)	63	78 (-)	79	78 (-)	82	65 (-)	70	50 (+)	47	75 (-)	79	79 (-)	86	64 (-)	70
NFBA	8 (+)	6	5 (+)	3	3	3	6 (+)	4	5 (+)	4	3 (+)	2	4 (+)	2	5 (+)	3	6 (+)	5	5 (+)	3	3 (+)	2	5 (+)	4
HHA	38 (-)	40	24 (+)	23	17 (-)	18	29 (+)	26	34 (+)	33	19	19	17 (+)	16	30 (+)	26	43 (-)	48	20 (+)	18	17 (+)	12	31 (+)	26
FA	1 (+)	0	0	0	0	0	0	0	0	0	0	0	0	0	0	0	1 (+)	0	0	0	0	0	1 (+)	0
Total	100	100	100	100	100	100	100	100	100	100	100	100	100	100	100	100	100	100	100	100	100	100	100	100

Note: LPR = low productivity states (Madhya Pradesh (MP), Rajasthan);
MPR = medium productivity states (Uttar Pradesh (UP), Bihar);
HPR = high productivity states (Punjab, Tamil Nadu).
Other definitions are as under Table 3.1.

Source: Estimated by authors.

Table 3.14

Disaggregative Analysis of Portfolio of Farm Business Assets: Three Agricultural Productivity Regions

Percentage Share

Components	LPR								MPR								HPR							
	S		M		L		Overall		S		M		L		Overall		S		M		L		Overall	
	II	I	II	I	II	I	II	I	II	I	II	I	II	I	II	I	II	I	II	I	II	I	II	I
L	90	90	91	91	92 (+)	88	91 (+)	90	93	93	91 (-)	92	95 (+)	92	92	92	93 (+)	91	93 (+)	90	94 (+)	93	93 (+)	91
B	3 (-)	4	2	2	1 (-)	2	2	2	3	3	2	2	1	1	3 (+)	2	3	3	1 (-)	2	1	1	2	2
AM and IC	4	4	5	5	6 (-)	8	5	5	3	3	6 (+)	5	4 (-)	6	4	4	3 (-)	4	4 (-)	6	4 (-)	5	4 (-)	6
AC	1 (-)	2	1	1	0 (-)	1	1	1	1 (-)	2	0 (-)	1	0	0	1	1	0 (-)	1	0	0	0	0	0	0
TE	1	1	1	1	1	1	1	1	0	0	1 (+)	0	0 (-)	1	0	0	1 (+)	0	1	1	1	1	1	1
Total	100	100	100	100	100	100	100	100	100	100	100	100	100	100	100	100	100	100	100	100	100	100	100	100

Note: Definitions of abbreviations are as under Tables 3.1 and 3.2.

Source: Estimated by authors.

Table 3.15

Disaggregative Analysis of Portfolio of Non-Farm Business Assets: Three Agricultural Productivity Regions

Percentage Share

Components	LPR								MPR								HPR							
	S		M		L		Overall		S		M		L		Overall		S		M		L		Overall	
	II	I	II	I	II	I	II	I	II	I	II	I	II	I	II	I	II	I	II	I	II	I	II	I
L	6	NA	1	NA	0	NA	4	NA	1	NA	0	NA	0	NA	1	NA	4	NA	32	NA	2	NA	16	NA
B	12 (+)	9	14 (+)	5	13 (-)	19	13 (+)	9	10 (-)	22	15 (+)	11	36 (-)	43	11 (-)	19	11 (-)	31	8 (-)	14	28 (-)	38	10 (-)	23
AC	67 (+)	60	75 (-)	80	74 (+)	62	70 (-)	79	78 (+)	52	74 (+)	67	24 (-)	31	77 (+)	55	64 (+)	31	40 (-)	46	49	49	53 (+)	39
NBM	9 (-)	15	4 (-)	6	3 (-)	11	7 (-)	10	5 (-)	20	6 (-)	10	4 (+)	3	5 (-)	17	8 (-)	27	14 (-)	27	17 (+)	0	11 (-)	26
TE	6 (-)	16	6 (-)	9	10 (+)	8	6 (-)	11	6	6	6 (-)	12	36 (+)	23	6 (-)	8	13 (+)	11	6 (-)	13	3 (-)	13	9 (-)	12
Total	100	100	100	100	100	100	100	100	100	100	100	100	100	100	100	100	100	100	100	100	100	100	100	100

Note: Definitions of abbreviations are as under Tables 3.1 and 3.3.

Source: Estimated by authors.

- In MPR, AC contributes the highest share (77%), followed by B asset (11%), TE (6%) and NBM (5%).
- In case of HPR, the highest share (53%) is contributed by AC, followed by L (16%), NBM (11%), B asset (10%) and TE (9%).

- The component analysis of HHA (Table 3.16) suggests the following major results:
  - In overall, the highest share, ranging between 60 per cent (MPR) and 50 per cent (HPR), is contributed by residential building asset, followed by land for household activities and household durable assets in all the regions.
  - Small farm households have experienced a rise in the share of residential buildings in all the regions, fall in the share of household durable assets and either a marginal rise or fall in the share of land asset.
  - In case of cultivator households of M category, there has been a rise in the share of residential buildings in LPR and MPR, but a fall in the share in HPR. Fall in the share of durable assets is observed in all the regions, whereas there has been a rise in the share of land and TE in case of this farm category.
  - With respect to large farm households, there has been a fall in the share of building and household durable assets, and rise in the share of land, as well as TE (except in MPR).

Growth Rate Analysis of Aggregative Groups of Assets

The growth rate analysis of only three broad groups of assets *viz.*, FBA, NFBA and HHA is performed with all-India total samples, and with data subsets of rural, urban and three regions of cultivator households. The growth rates between 1991-92 and 2002-03 are computed. The growth rates of financial assets are not computed in view of the almost zero/insignificant value of these assets. The results on growth rates are given in Tables 3.17-3.20.

At all-India level, the highest growth rate (9.3%) is observed in case of NFBA, followed by HHA (6.4%) and FBA (4.7%). The implication of this is that non-farm business has been growing at a higher rate, and less

Table 3.16

Disaggregative Analysis of Portfolio of Household Assets: Three Agricultural Productivity Regions

Percentage Share

Components	LPR								MPR								HPR							
	S		M		L		Overall		S		M		L		Overall		S		M		L		Overall	
	II	I	II	I	II	I	II	I	II	I	II	I	II	I	II	I	II	I	II	I	II	I	II	I
L	26 (-)	28	26 (+)	24	26 (+)	18	26 (+)	24	29 (+)	28	28 (+)	26	52 (+)	27	29 (+)	28	31 (-)	33	39 (+)	30	34 (+)	24	33 (+)	31
B	59 (+)	55	57 (+)	55	55 (-)	56	58 (+)	55	60 (+)	58	58 (+)	57	37 (-)	51	60 (+)	58	52 (+)	47	45 (-)	51	44 (-)	53	50 (+)	49
TE	1 (-)	2	3 (+)	2	4 (+)	3	2	2	1	1	3 (+)	2	2 (-)	5	2 (+)	1	2 (+)	1	3 (+)	2	6 (+)	4	2	2
DG	13 (-)	15	14 (-)	20	16 (-)	23	14 (-)	19	9 (-)	12	11 (-)	16	9 (-)	17	9 (-)	14	15 (-)	19	13 (-)	17	16 (-)	20	14 (-)	18
Total	100	100	100	100	100	100	100	100	100	100	100	100	100	100	100	100	100	100	100	100	100	100	100	100

Note: Definitions abbreviations are as under Tables 3.1 and 3.4.

Source: Estimated by authors.

importance has been given to accumulate FBA (Table 3.17). Further, the growth rates of NFBA, FBA and HHA are found to be higher among S category of farm households, followed by growth rates in case of M category of farm households. These growth rates are far below in case of L farms.

With respect to rural farm households (Table 3.18), the growth patterns are the same as that of pooled data set of cultivator households (Table 3.17).

Table 3.17

Growth Rates of Assets: All India

Asset Group	*S*	*M*	*L*	*Overall*
FBA	8.1	2.3	1.4	4.7
NFBA	11.0	7.1	2.8	9.3
HHA	8.2	3.0	0.6	6.4

Note: Definitions of abbreviations are as in Table 3.1.
Source: Estimated by authors.

Table 3.18

Growth Rates of Assets: Rural Households

Asset Group	*S*	*M*	*L*	*Overall*
FBA	8.4	1.9	0.3	4.6
NFBA	14.6	7.0	1.9	11.3
HHA	9.7	2.1	-1.0	7.1

Note: Definitions of abbreviations are as in Table 3.1

The asset growth in urban cultivator households unfolds a different pattern (Table 3.19), compared to those of rural cultivator households. The growth rate (6.1%) of FBA is the highest followed by HHA and NFBA. The same pattern of growth rates is observed in case of three farm groups. Further, the urban M category of households are found to be moving towards accumulating more NFBA, whereas small farm households towards accumulating more of FBA, compared to their accumulation of NFBA and HHA.

Table 3.19

Growth Rates of Assets: Urban Households

Asset Group	*S*	*M*	*L*	*Overall*
FBA	5.3	6.6	7.5	6.1
NFBA	1.9	7.2	4.3	3.5
HHA	3.5	7.1	3.9	4.3

Note: Definitions of abbreviations are as in Table 3.1.

Source: Estimated by authors.

The asset growth pattern analysis of farm households in three regions would point towards the following inferences (Table 3.20):

- In overall, the farm households of LPR are observed to have experienced highest growth rate (10.6%) of NFBA, followed by households of MPR (9%) and of HPR (8%). This is an expected pattern of asset growth, because households of LPR have to increase their income from non-farm activity by accumulation of NFBA. However, in all the regions, the highest growth rates of NFBA have been registered compared to those of FBA and HHA.
- The total asset growth rate appears to be the highest in LPR, followed by growth rates in MPR and HPR. In other words, HPR is in the trajectory of low growth rate of pooled total assets *viz.*, FBA, NFBA and HHA, whereas LPR is in the trajectory of high growth rate.

Table 3.20

Growth Rates of Assets: Three Agricultural Productivity Regions

Asset Group	*LPR*				*MPR*				*HPR*			
	S	*M*	*L*	*Overall*	*S*	*M*	*L*	*Overall*	*S*	*M*	*L*	*Overall*
FBA	11.6	4.7	0.3	5.6	8.2	0.1	-4.3	4.7	8.9	1.9	-0.1	3.6
NFBA	14.8	8.8	-0.1	10.6	10.5	3.6	2.3	9.0	9.5	6.8	5.3	8.0
HHA	11.0	5.3	0.0	7.4	8.9	0.2	-3.3	6.7	7.4	3.7	4.1	6.0

Note: Definitions of abbreviations are as in Table 3.1.

Source: Estimated by authors.

- Yet another interesting inference is that small farmers are found to have registered the highest growth rate of total assets, in all the regions followed by medium and large farmers. Further in case of small farmers, the growth rate of NFBA is the highest in all the regions than that of FBA and HHA.

The implications of those patterns of asset growth will be discussed in Chapter 7 to derive implications for policy and programme designs.

Assets of Cultivator and Non-cultivator Households in Rural India: Temporal Variations in Composition and Growth

In addition to compositional aggregative and disaggregative analysis of assets structure of cultivator households—both rural and urban—during 1991-92 and 2002-03 under the section Aggregative and Disaggregative Analysis of Assets Structure, yet another way is to examine the dynamics of structure of assets accumulated by rural cultivator and non-cultivator households during 1971-2002, forming three decades of empirical evidences. It needs to be kept in mind, the definition of assets especially in case of land, building and machinery and equipment is different here compared to the ones used in the section: Aggregative and Disaggregative Analysis of Assets Structure, and one addition made to the asset group is 'Dues Receivable.' Added to this is the point that the asset structure discussed in this section is also based on unit-level (household level) data collected by NSSO. The results are given in Tables 3.21 and 3.22.

The three decades of empirical evidences on asset structure of rural cultivator and non-cultivator households are given in Table 3.21. Six major inferences are extracted from these results:

- In case of cultivator households, more than two-third of assets have been in the form of land, and this percentage share has almost remained constant over a period of time. On the other hand, both land and building are found to have contributed over one-third each to the total assets held by non-cultivator households. Excepting some marginal shifts in percentage share, the shares of these two assets have remained almost constant over a period of three decades.

- With respect to livestock and poultry, the percentage share in total assets held by cultivator and non-cultivator households was almost the same during 1971 and 1981. But during 1991 and 2002, the non-cultivator households have experienced a greater decline in the share of this asset. However, both types of households are found to have registered a considerable fall in the share of livestock and poultry in their total assets since 1991.

Table 3.21

Share of Different Assets in Cultivator and Non-Cultivator Households in Rural India during the Last Three Decades: 1971-2002

Percentages

Asset	*Cultivator Households*				*Non-Cultivator Households*			
	1971	*1981*	*1991*	*2002*	*1971*	*1981*	*1991*	*2002*
Land	69.0	64.1	67.6	68.1	32.3	30.8	39.8	38.2
Building	16.5	19.5	19.1	20.1	39.7	39.1	37.9	41.4
Livestock and Poultry	6.5	5.0	3.4	2.3	6.4	5.2	2.9	1.3
Machinery and Equipment	2.7	3.7	3.8	3.8	3.1	3.3	3.5	3.4
Household Durables	4.2	6.6	5.1	4.2	11.5	15.8	11.2	10.0
Financial Assets	0.8	1.0	0.8	1.6	5.2	5.5	4.6	5.4
Dues Receivable	0.6	0.1	0.1	0.1	1.8	0.3	0.1	0.2
Total	100.0	100.0	100.0	100.0	100.0	100.0	100.0	100.0

Note: Definitions of assets:

- *Land*: Total land area owned including orchards and plantations area put to non-agricultural uses and residential area.
- *Building*: Residential building, building used for farm business, building used for non-farm business.
- *Livestock and Poultry*: Livestock (cattle, sheep, goat etc), fishing, beekeeping and dairy farming.
- *Machinery and Equipment:* Agricultural machinery, non-farm business equipment, transport equipment.
- *Household Durables:* Major infrequently purchased goods for domestic use, and had a long expected life.
- *Financial Assets:* Shares and debentures, certificates, deposits in post office, cooperatives, commercial banks, non-banking company, chit contribution, insurance premium, annuity certificates, cash on hand, deposit with individual and other financial assets.
- *Dues Receivable:* Promissory note, mortgage of real estate, pledge of ornaments/other movable property, unsecured loan, professional dues, trade credit, kind loans and others.

Source: Developed from unit level data of NSSO, Government of India.

- The percentage share of machinery and equipment has remained less than 4 per cent of total assets over a period of three decades in both cultivator and non-cultivator household. But the concept of machinery and equipment used in this analysis consists those of farm business and non-farm business, including transport equipment. Further, in the case of non-cultivator households, this asset is for non-farm business activities.

- The percentage share of household durable assets in total assets is far higher in non-cultivator households than in cultivator households—a little less than three-fold higher.
- Compared to cultivator households, the non-cultivator households have revealed their preference for holding more than 5 per cent of their assets in the form of financial assets, whereas the percentage share of financial assets in cultivator households is quite insignificant, bordering in zero share in total assets. The difference is a measure of difference in the participation of these two types of households in capital and money markets.
- With respect to dues receivable also, the non-cultivator households are found to be ahead of cultivator households, even though the percentage share of dues receivable is insignificant in total assets.

The unit-level data pool of NSSO for the period 1991 and 2002 is employed for analysing the shares of different assets in cultivator and non-cultivator household of six states of high (HPR), medium (MPR) and low (LPR) agricultural productivity regions in the country (Table 3.22).

- Asset Preference Structure of Cultivator Households:

  Between two states *viz.*, Punjab and Tamil Nadu (TN) of HPR, one could observe quite a contrast in the percentage shares of land and building in total assets of cultivator households. In Punjab, land is found to be a more dominant asset than in TN, and in TN, the share of building is almost double compared to what it is in Punjab. However, the share of land as well as building in both the states is found to be over 80 per cent of total assets.

  The percentage share of livestock and pountry in both the states has declined over a period of time. Further, with respect to financial assets, the revealed preference of TN cultivator households has been more than what it is in Punjab.

  - With regard to cultivator households of MPR states, over 80 per cent of the share in total assets is in the form of land

Table 3.22

Share of Different Assets in Cultivator and Non-Cultivator Households in Rural India during 1991 and 2002 in the States of Three Agricultural Productivity Regions

Percentages

Productivity Regions/States	*Land*		*Building*		*Livestock/ Poultry*		*Machinery/ Equipment*		*HHDA*		*Financial Assets*		*Loan Receivable*	
	I	*II*	*I*	*II*	*I*	*II*	*I*	*II*	*I*	*II*	*I*	*II*	*I*	*II*
Assets: Cultivator Households														
HPR:														
Punjab	77.9	75.9	11.8	13.5	1.9	1.5	5.8	5.4	2.2	2.7	0.4	0.9	0.0	0.1
Tamil Nadu	57.6	58.8	23.0	25.2	2.6	1.7	4.7	3.8	10.3	8.2	1.6	2.1	0.2	0.2
MPR:														
Uttar Pradesh	70.3	67.1	18.3	21.6	3.1	2.2	3.9	3.3	3.9	3.4	0.5	2.4	0.0	0.0
Bihar	73.5	66.1	17.7	22.9	3.0	3.2	2.0	3.0	3.5	3.5	0.4	1.3	0.0	0.1
LPR:														
Madhya Pradesh	71.8	69.4	15.0	18.2	4.8	2.9	4.4	5.0	3.4	3.4	0.5	1.1	0.0	0.0
Rajasthan	64.6	66.7	19.6	20.3	4.1	2.8	4.6	4.3	6.5	4.9	0.6	0.9	0.0	0.1
Assets: Non-Cultivator Households														
HPR:														
Punjab	49.3	46.2	34.8	36.1	2.8	1.1	3.0	3.3	7.5	8.2	2.6	4.8	0.1	0.3
Tamil Nadu	24.8	25.1	46.5	47.7	1.6	0.6	4.6	3.5	18.7	16.4	3.6	6.5	0.1	0.3
MPR:														
Uttar Pradesh	50.8	51.3	34.8	35.7	3.2	1.2	3.4	1.6	6.2	4.6	1.6	5.6	0.1	0.0
Bihar	41.0	29.4	40.6	45.4	3.6	1.8	2.9	3.9	9.1	8.2	2.7	11.2	0.0	0.1
LPR:														
Madhya Pradesh	45.2	34.3	33.5	43.2	4.3	1.4	3.4	4.6	8.7	8.9	4.9	7.6	0.0	0.1
Rajasthan	46.7	47.3	34.7	33.7	2.6	1.5	3.3	4.7	11.0	9.5	1.7	3.1	0.0	0.3

Note: Definitions of assets and source of data are as in Table 3.21.
I = 1991 II = 2002.

Source: Estimated by authors.

and building in Uttar Pradesh (UP) as well as Bihar. UP has experienced a fall in the share of livestock and poultry, and machinery and equipment, whereas Bihar has experienced an increase in the share of these two assets. In case of performance for financial assets, the cultivator households of UP are ahead of Bihar households.

- In the two states of LPR, land is still the dominant asset, followed by building. The share of livestock and poultry has declined in both the states over a period of time. The share of machinery and equipment has remained a little over 4 per cent. The cultivator households of Rajasthan are found to have a greater share of household assets in their total assets. With respect of financial assets, the households of both the states have experienced some increase in the share out of their total assets.

- Asset Preference Structure of Non-Cultivator Households:
  - In HPR, compared to non-cultivator households of Punjab, those of TN have higher preference for building, for machinery and equipment, for household assets and for financial assets.
  - In MPR, the households of UP have higher preference for land, whereas in Bihar almost the first preference is for buildings. But regarding the share of land and buildings considered together, over 80 per cent of preference is in favour of these two assets in both the states. In both the states, there has been a considerable decline in the share of livestock and poultry. The preference of non-cultivator households of Bihar is more in favour of household assets and financial assets.
  - In the states of LPR, both land and building are the dominant assets, forming around 80 per cent of total assets. In case of livestock and poultry assets, there has been a considerable decline in the share over a period of one decade. The share of machinery and equipment has increased in both MP and Rajasthan. With respect to the share of financial assets in total assets, there has been considerable increase over a period of time, but MP is found to be ahead of Rajasthan.

What Inferences could be Drawn on Inter-regional Variations and Similarities in the Composition of Assets in both Cultivator and Non-cultivator Households?:

- In case of cultivator households of all three regions/six states: First, the share of land and building forms over 80 per cent of total assets. Second, the percentage share of livestock and poultry has declined over a period of time in all the six states. Third, even though the share of financial assets is quite low in all the states, there has been a marginal increase in the share of this asset. Fourth, excepting some cases, there has been a fall in the share of machinery and equipment in the study states, implying less of non-land asset available for agriculture.
- In case of non-cultivator households: First, land and building together constitute the dominant asset group in all the regions. Second, as in the case of cultivator households, there has been a decline in the share of livestock and poultry. Third, in majority of states, there has been an increase in the share of machinery and equipment, obviously for non-farm business activities. Fourth, in all the study states, there has been a considerable increase in the share of financial assets, and more over, the share of financial assets in non-cultivator households is far higher than in those of cultivator households. Some non-cultivator households are likely to be traders-cum-money lenders.

Asset Growth Rate Patterns in Cultivator and Non-Cultivator Households of Rural India and Regions

In addition to the analysis of growth rates of farm and non-farm business assets, and household assets in the preceding section of this chapter, an analysis of asset growth rates (between 1991-92 and 2002-03) of seven asset groups as defined in Table 3.21 both at all-India and at regional level is in order. The results on this are given in Table 3.23.

- Estimates with all-India sample suggest that:
  - The overall growth rate of all seven assets in non-cultivator households is found to be higher than that of cultivator households. This higher overall growth rate in non-cultivator

Table 3.23

Annual Growth Rates (%) of Assets in Rural Cultivator and Non-Cultivator Households between 1991-92 and 2002-03: All India and in the States of Three Agricultural Productivity Regions

Assets	*All India*		*Regions/States*											
			HPR				*MPR*				*LPR*			
			TN		*Punjab*		*UP*		*Bihar*		*MP*		*Rajasthan*	
	Cult.	*Non-Cult.*	*Cult.*	*Non-Cult.*	*Cult.*	*Non-Cult.*	*Cult.*	*Non-Cult.*	*Cult.*	*Non-Cult.*	*Cult.*	*Non-Cult.*	*Cult.*	*Non-Cult.*
1. Land	1.20	1.41	3.20	3.71	0.01	-0.24	7.17	8.61	3.37	3.60	6.28	3.37	0.08	3.41
2. Building	1.57	2.62	3.87	3.84	1.49	0.70	9.29	8.77	6.86	7.86	8.50	8.48	0.07	3.02
3. Livestock and Poultry	-0.30	-0.61	-0.10	-0.66	-0.21	-0.90	0.47	-0.06	0.52	0.05	0.18	-0.51	-0.38	-0.24
4. Machinery and Equipment	1.09	1.51	1.10	1.04	-0.36	1.24	5.81	1.49	8.45	9.72	7.81	8.88	-0.89	6.64
5. Durable Household Assets	-0.82	0.75	0.96	2.36	2.29	1.25	6.51	5.50	4.44	5.69	6.49	6.23	-2.27	1.97
6. Financial Assets	0.80	0.37	0.57	0.98	0.76	0.65	2.33	2.15	1.62	2.17	1.53	1.08	0.43	0.97
7. Loans Receivable	6.99	3.36	1.17	12.64	17.30	14.52	13.01	-16.15	13.75	19.36	3.02	22.21	14.71	31.22
Overall Growth Rates	1.14	1.78	3.01	3.61	0.25	0.35	7.63	8.51	4.37	6.78	6.61	6.00	-0.22	3.29

Note: Definitions are as in Table 3.21.

Source: Estimated from NSSO unit-level data for the years 1991-92 to 2002-03.

households could be attributed to higher growth rates of assets such as land, building, machinery and equipment, and durable household assets of these non-cultivator households.

- Estimates with sample of the six states of three agricultural productivity regions suggest that:
  - In HPR, the overall growth rates of all the seven assets are found to be higher in non-cultivator households of both TN and Punjab. Further the overall growth rates of assets in both cultivator and non-cultivator households of TN are far higher than those of Punjab. Perhaps, this higher asset growth performance of TN could be attributed to rather high growth rates of two assets *viz.*, land and building, whereas in Punjab, the growth rates of these two assets are far below.
  - With respect to MPR, both the cultivator and non-cultivator households of UP have experienced higher growth rates than of Bihar. This overall higher growth rates of UP could be attributed mainly to higher growth rates of land and building assets in the state. However the overall growth rates of non-cultivator households of both the States are quite higher than those of cultivator households. Added to this pattern of growth are the high growth rates of household durable assets and low growth rates of livestock and poultry in both the states of cultivator and non-cultivator households.
  - With respect to the states of LPR, the overall asset growth rates in both cultivator and non-cultivator households of MP are quite higher than those of Rajasthan. This could be attributed mainly to high growth rates of land, building, machinery and equipment, and durable household goods. Only in loans (Dues) receivable, the households of Rajasthan are found to be far ahead. Further between cultivator and non-cultivator households, the non-cultivator households of both the states are ahead of cultivator households with respect to the asset category of 'loans receivable.'

Determinants of Accumulation of Non-Farm Business Assets: Rural and Urban Farm Households

Six major concepts and evidences derived from cross-country experiences discussed in the first section of this chapter are recalled to place the empirical evidences of this study on determinants of non-farm business assets in proper perspective. First, the degree of vulnerability and insecurity of households depend heavily on asset ownership. For this reason, asset-based social policy is treated as a cutting edge development agenda. Second, policy designs are needed to 'influence the pattern of asset portfolio, growth and determinants.' Third, the well-being of a household depends on the stock of assets and its composition. Farm households have multiple forms of assets (such as farm business assets, non-farm business assets, household durable assets and financial assets) and liabilities. Fourth, asset, activity and income diversification are said to lie at the heart of livelihood strategies of poor communities. Fifth, not much is known about non-farm business assets held by farmers for non-farm activities such as bicycle repair, transport, flour mill, carpentry, sale of cooked food, livestock rearing (like poultry and piggery) and so on. There are some empirical regularities to indicate positive relationship between investment in non-farm business assets, and income as well as household welfare across most of the rural Africa. Sixth, it is possible to postulate the determinants of investment in non-farm business assets such as age, education, net worth, family size and many other variables.

Further, it is also recalled from from the section: All-India Aggregative and Disaggregative Results that: (a) over a period of time, there has been a gain in the share of non-business assets in total assets owned by farm households. This result is suggestive of the inference that cultivator households have made a development move, however small the move is towards accumulation of non-farm business category of assets. It could be considered as a symptom of transition towards non-farm activity to reduce the variance in their income and consumption levels. (b) Some of the components of non-farm business assets are animal capital, building, machinery and equipment (including transport equipment) for non-farm business activities.

Against the backdrop of the section preceded, the results on determinants of non-farm business assets of rural and urban households are examined. It is postulated that education of the household head, age of the household head, net worth (the difference between total assets and total liabilities), farm size, and household size are the major determinants of non-farm business asset accumulation. Further, it is also postulated that there are regional effects. The empirical results are given in Tables 3.24-3.27.

It is observed from Table 3.24 (Rural) that net worth, farm size and household size have positive and significant impact on 'accumulation' of non-farm business assets by rural farm households during both the first and the second period. Age of the head has positive and significant impact only during the second period. But surprisingly literacy is found to have negative and significant impact on the same dependent variable compared to illiteracy during the first period.

Table 3.24

Ordinary Least Squares (OLS) Regression of Non-Farm Business Asset: All-India: Rural

	Elasticities/ Coefficients		*Standard Error*		*t-value*		*P-value*	
Variables	*II*	*I*	*II*	*I*	*II*	*I*	*II*	*I*
AoH	0.154	-0.048	0.035	0.039	4.35	-1.24	***	
DEDU	-0.023	-0.131	0.020	0.025	-1.18	-5.33		***
NW	0.533	0.539	0.010	0.013	52.43	41.48	***	***
FS	0.601	0.586	0.040	0.042	15.18	13.97	***	***
HHS	0.999	0.829	0.021	0.025	48.04	33.56	***	***
Constant	-2.462	-2.077	0.128	0.167	-19.2	-12.44	***	***
Adjusted R Square	0.2051	0.2305						
Number of Sample	45054	21700						

Note: AoH = Log Function of Age of Head;
DEDU = Dummy variable of literacy/illiteracy of head where literacy = 1 and illiteracy = 0;
NW = Log Function of Net Worth;
FS = Log Function of Farm Size (Total assets – Total outstanding debt);
HHS = Log Function of Household Size.

Source: Estimated by authors.

The regression results (Table 3.25) derived from the sample of urban farm households suggest that during both the periods, net worth, farm size and household size have positive and significant impact on non-farm business assets. Surprisingly, literacy has negative and significant impact during both the periods.

The estimated regression coefficients (elasticities) of non-farm business assets (Tables 3.24 and 3.25) derived from the rural and urban sample households of three regions suggest that:

- During both the periods, age of the head, net worth, farm size and household size have positive and significant impact on non-farm business asset accumulation in case of rural households. The coefficients of dummy variables are negative and significant for HPR and MPR (Table 3.26).

Table 3.25

OLS Regression of Non-Farm Business Asset: All-India: Urban

Variables	*Elasticities/ Coefficients*		*Standard Error*		*t-value*		*P-value*	
	II	*I*	*II*	*I*	*II*	*I*	*II*	*I*
AoH	-0.214	-0.147	0.165	0.155	-1.3	-0.94		
DEDU	-0.590	-0.384	0.102	0.105	-5.77	-3.66	***	***
NW	0.502	0.500	0.041	0.058	12.13	8.69	***	***
FS	0.416	0.855	0.186	0.217	2.24	3.93	**	***
HHS	1.067	1.069	0.094	0.102	11.36	10.51	***	***
Constant	-0.931	-2.144	0.647	0.630	-1.44	-3.4		***
Adjusted R Square	0.1276	0.1635						
Number of Sample	3241	2335						

Source: Estimated by authors.

Table 3.26

OLS Regression of Non-Farm Business Asset: Three Regions: Rural

Variables	*Elasticities/ Coefficients*		*Standard Error*		*t-value*		*P-value*	
	II	*I*	*II*	*I*	*II*	*I*	*II*	*I*
AoH	0.334	0.118	0.039	0.051	8.52	2.32	***	**
DEDU	-0.019	-0.145	0.022	0.034	-0.85	-4.3		***
NW	0.550	0.539	0.013	0.020	41.19	27.32	***	***
FS	0.367	0.570	0.049	0.073	7.44	7.77	***	***
HHS	0.919	0.788	0.024	0.033	37.83	23.88	***	***
DHPR	-0.283	-0.366	0.045	0.057	-6.23	-6.47	***	***
DMPR	-0.359	-0.314	0.024	0.044	-14.87	-7.23	***	***
Constant	-2.825	-2.283	0.171	0.211	-16.49	-10.82	***	***
Adjusted R Square	0.2205	0.2609						
Number of Sample	24707	11126						

Note: DHPR = Dummy Variable for High Productivity Region;
DMPR = Dummy Variable for Medium Productivity Region.

Source: Estimated by authors.

- In case of urban samples (Table 3.27), only two variables *viz.*, net worth and household size are found to have had positive and significant impact on non-farm business assets during both second and the first period. Farm size has exerted positive and significant impact on dependent variable during the first period. The coefficient of dummy variable is lower for MPR than for LPR during both the periods.

Table 3.27

OLS Regression of Non-Farm Business Asset: Three Regions: Urban

	Elasticities/ Coefficients		*Standard Error*		*t-value*		*P-value*	
Variables	*II*	*I*	*II*	*I*	*II*	*I*	*II*	*I*
AoH	0.021	-0.083	0.203	0.239	0.1	-0.35		
DEDU	-0.577	-0.264	0.126	0.129	-4.57	-2.04	***	**
NW	0.544	0.603	0.058	0.084	9.42	7.15	***	***
FS	0.223	0.565	0.241	0.309	0.93	1.83		*
HHS	0.948	1.224	0.115	0.120	8.25	10.23	***	***
DHPR	-0.233	-0.238	0.161	0.195	-1.45	-1.22		
DMPR	-0.181	-0.340	0.112	0.191	-1.62	-1.78	***	*
Constant	-1.741	-3.011	0.825	0.976	-2.11	-3.09	**	***
Adjusted R Square	0.1437	0.2102						
Number of Sample	1735	1002						

Source: Estimated by authors.

Determinants of Total Non-Land Assets

An identification as well as analysis of determinants of total non-land assets would yet form an important dimension of asset portfolio analysis. The empirical results with all-India rural and urban sample and rural and urban samples of three agriculture productivity regions are given in Tables 3.28-3.31. It is postulated that education level and age of the household head, household size, total expenditure incurred by the household and farm size are the major determinants of total non-land farm assets.

It could be observed from Table 3.28 that all the coefficients (elasticities) for both P.I and P.II are both positive and statistically significant. Literacy and age of the head of the household, household size, total expenditure and farm size are found to have positive impact on formation of total non-land capital in rural households. The same inference holds good in case of urban sample households (Table 3.29).

Table 3.28

OLS Regression of Total Non-Land Assets: All-India: Rural

	Elasticities/ Coefficients		*Standard Error*		*t-value*		*P-value*	
Variables	*II*	*I*	*II*	*I*	*II*	*I*	*II*	*I*
EoH	0.355	0.553	0.006	0.012	60.56	45.55	***	***
AoH	0.720	0.694	0.017	0.033	42.64	21.24	***	***
HHS	0.359	0.374	0.008	0.014	43.36	26.2	***	***
EXP	0.096	0.071	0.002	0.003	45.15	22.64	***	***
FS	0.777	0.789	0.015	0.019	52.72	42.14	***	***
Constant	2.972	2.903	0.066	0.122	45.03	23.75	***	***
Adjusted R Square	0.2932	0.2990						
Number of Sample	45054	21700						

Note: EoH = Log Function of Education of Head;
AoH = Log Function of Age of Head;
HHS = Log Function of Household Size;
EXP = Log Function of Total Expenditure;
FS = Log Function of Farm Size.

Source: Estimated by authors.

Table 3.29

OLS Regression of Total Non-Land Assets: All-India: Urban

	Elasticities/ Coefficients		*Standard Error*		*t-value*		*P-value*	
Variables	*II*	*I*	*II*	*I*	*II*	*I*	*II*	*I*
EoH	0.503	0.809	0.022	0.039	22.58	20.55	***	***
AoH	1.130	0.969	0.070	0.138	16.21	7.02	***	***
HHS	0.262	0.276	0.038	0.053	6.99	5.22	***	***
EXP	0.113	0.105	0.008	0.009	14.14	11.98	***	***
FS	0.656	0.592	0.058	0.075	11.22	7.91	***	***
Constant	2.039	2.408	0.285	0.531	7.16	4.54	***	***
Adjusted R Square	0.3073	0.2896						
Number of Sample	3241	2335						

Source: Estimated by authors.

As could be seen from Tables 3.30 and 3.31, the impact of all the explanatory variables (determinants) is found to be positive on total non-land assets in case of rural and urban samples from three agricultural productivity regions. All the elasticities are statistically significant during both the periods, excepting the elasticities of farm size during both the periods. However, the coefficients of farm size are positive. Further, compared to LPR, both in HPR and MPR, the coefficients are not merely positive but also statistically significant in case of both rural and urban samples of both the periods. These are suggestive of the inference that there is a positive association between agricultural productivity and non-land capital stock.

Table 3.30

OLS Regression of Total Non-Land Assets: Three Regions on Agricultural Productivity: Rural

	Elasticities/ Coefficients		*Standard Error*		*t-value*		*P-value*	
Variables	*II*	*I*	*II*	*I*	*II*	*I*	*II*	*I*
EoH	0.281	0.452	0.006	0.015	48.27	30.16	***	***
AoH	0.535	0.579	0.020	0.040	26.23	14.48	***	***
HHS	0.416	0.370	0.010	0.017	42.29	21.12	***	***
EXP	0.091	0.071	0.003	0.004	32.24	15.74	***	***
DHPR	0.874	0.947	0.021	0.031	40.77	30.41	***	***
DMPR	0.500	0.436	0.019	0.034	26.2	12.68	***	***
FS	0.098	0.261	0.011	0.023	8.71	11.37	***	***
Constant	3.553	3.230	0.077	0.146	46.21	22.16	***	***
Adjusted R Square	0.3351	0.3380						
Number of Sample	24707	11126						

Note: DHPR = Dummy Variable for High Productivity Region.
DMPR = Dummy Variable for Medium Productivity Region. Low Productivity Region is taken as base.

Source: Estimated by authors.

Table 3.31

OLS Regression of Total Non-land Assets: Three Regions on Agricultural Productivity: Urban

	Elasticities/ Coefficients		*Standard Error*		*t-value*		*P-value*	
Variables	*II*	*I*	*II*	*I*	*II*	*I*	*II*	*I*
EoH	0.413	0.680	0.026	0.053	15.98	12.79	***	***
AoH	0.745	0.770	0.075	0.112	9.92	6.89	***	***
HHS	0.407	0.299	0.043	0.060	9.48	4.99	***	***
EXP	0.114	0.131	0.011	0.017	10.8	7.66	***	***
DHPR	0.896	0.763	0.075	0.097	11.96	7.87	***	***
DMPR	0.434	0.374	0.058	0.107	7.47	3.5	***	**
FS	0.056	0.076	0.049	0.069	1.15	1.09		
Constant	3.106	3.046	0.309	0.413	10.04	7.37	***	***
Adjusted R Square	0.3679	0.3304						
Number of Sample	1735	1002						

Source: Estimated by authors.

4 Household Liability Analysis

Debt Portfolio and Asset-Debt Relations

In this chapter, cultivator household liability is the main focus of examination with results from aggregative as well as disaggregative data sets. There are nine major issues examined around household liability analysis. First, how is the composition of liability (debt) as between institutional and non-institutional agencies? Second, for which broad group of asset (ex. farm business asset) accumulation, the households have incurred debt liability? Third, how are the debt-asset relationships with respect to debt from institutional and non-institutional agencies, and with respect to different activities like that of farm business and non-farm business activities? Fourth, what are the determinants of household debt-asset ratio? Fifth, what factors would govern the intensity of financial inclusion? Sixth, what does the temporal analysis of financial exclusion and inclusion of rural households suggest? How are credit availment and asset holding related? Seventh, how are the temporal trends in the flow of production and investment credit to agriculture sector, and what is the pattern of investment credit distribution? Eighth, what is the relationship between asset portfolio distribution and credit availment across farm groups? Ninth, what is the pattern of relationship between FBA and credit availment from two broad types of agencies *viz.*, institutional and non-institutional? All these issues could be examined in the framework of the development premise that easily available financial services are treated as a key requirement for income growth. This has been established by both theoretical and empirical investigations (Rajan and Zingales, 1998; Hanohan, 2004; Levine, 2005; Government of India, 2008).

All these major issues are examined with results from all-India pooled data, rural and urban cultivator household data, and from three regions (low-productivity, LPR; medium-productivity, MPR; and high-productivity,

HPR) and other related sources. Farm size-wise results are also analysed from these aggregative and disaggregative data sets.

Agency-wise Composition of Household Liability

In Tables 4.1-4.4, the agency-wise (institutional and non-institutional) compositions of household outstanding loans are presented, across farm groups at all India, rural and urban, and at three regions.

At all-India level (cutting across rural and urban, and farm groups), 61 per cent of outstanding loans are from institutional sources, and 39 per cent from non-institutional sources during the second period. Further, the share of institutional loans in total outstanding loans has declined and that of non-institutional loans increased. However small it is, there has been a trend towards more of financial exclusion of farmers from institutional sources over a period of time.

Farm group-wise, medium (M) and large (L) farms are found to have increased the share from institutional sources, much more so in case of L farms. On the other hand, small (S) farm households are found to have experienced less of access to institutional sources (in terms of decrease in the share of institutional loans out of total outstanding loans) over a period of time (Table 4.1).

Table 4.1

Compositions of Outstanding Loans and Temporal Shifts: All India

Percentage Share

	S		*M*		*L*		*Overall*	
	II	*I*	*II*	*I*	*II*	*I*	*II*	*I*
Institutional Loans	55 (-)	63	69 (+)	68	74 (+)	66	61 (-)	64
Non-institutional Loans	45 (+)	37	31 (-)	32	26 (-)	34	39 (+)	36
Total	100	100	100	100	100	100	100	100

Note: I = 1991-92, II = 2002-03.
S = marginal and small farms.
M = semi-medium and medium farms.
L = large farms.

Source: Estimated by authors.

In case of rural cultivator households, there is marginal difference in the percentage share of institutional and non-institutional sources of credit availed by rural farm households between the two periods of time. But farm-wise, the M farms have increased their share from institutional sources, whereas S farm households have experienced a fall in the share of institutional loans out of total outstanding loans (Table 4.2).

Table 4.2

Compositions of Outstanding Loans and Temporal Shifts: All India: Rural

Percentage Share

	S		*M*		*L*		*Overall*	
	II	*I*	*II*	*I*	*II*	*I*	*II*	*I*
Institutional Loans	53 (-)	60	69 (+)	65	69	69	59 (-)	60
Non-institutional Loans	47 (+)	40	31 (-)	35	31	31	41 (+)	40
Total	100	100	100	100	100	100	100	100

Source: Estimated by authors.

In general, the urban households have recorded the experience of higher share of institutional sources (between 80 and 70%), compared to rural households in case whom the share ranges between 60 and 59 per cent. But in case of urban farm household also, there has been a considerable decline in the share of institutional loans, and obviously an increase in the share of non-institutional sources over a period of time.

With respect to the share of institutional loans in total outstanding loans, L farm groups have increased their share to 96 per cent, whereas M and S farm groups have experienced a fall in the share of institutional loans (Table 4.3).

Table 4.3

Compositions of Outstanding Loans and Temporal Shifts: All India: Urban

Percentage Share

	S		*M*		*L*		*Overall*	
	II	*I*	*II*	*I*	*II*	*I*	*II*	*I*
Institutional Loans	68 (-)	69	71 (-)	83	96 (+)	58	70 (-)	83
Non-institutional Loans	32 (+)	31	29 (+)	17	4 (-)	42	30 (+)	17
Total	100	100	100	100	100	100	100	100

Source: Estimated by authors.

Table 4.4

Compositions of Outstanding Loans and Temporal Shifts in Three Regions

Percentage Share

	LPR								MPR								HPR							
	S		M		L		Overall		S		M		L		Overall		S		M		L		Overall	
	II	I	II	I	II	I	II	I	II	I	II	I	II	I	II	I	II	I	II	I	II	I	II	I
IL	42 (-)	53	63 (+)	62	65 (-)	69	53 (-)	60	56 (-)	72	90 (+)	88	80 (+)	31	64 (-)	79	49 (-)	53	61 (-)	64	93 (+)	73	56 (-)	59
NIL	58 (+)	47	37 (-)	38	35 (+)	31	47 (+)	40	44 (+)	28	10 (-)	12	20 (-)	69	36 (+)	21	51 (+)	47	39 (+)	36	7 (-)	27	44 (+)	41
Total	100	100	100	100	100	100	100	100	100	100	100	100	100	100	100	100	100	100	100	100	100	100	100	100

Note: IL = Institutional loans.
NIL = Non-institutional loans.

Source: Estimated by authors.

The overall results on the share of institutional and non-institutional loan in total outstanding loan in three regions suggest (Table 4.4) that the percentage share of institutional loans has declined over a period of time, the decline being the highest in MPR and lowest in HPR. Compared to the share in the first period, the S farm groups have experienced a fall in the share of institutional loans during the second period in all the regions. The same trend is found in case of M farms of HPR, whereas this farm group has experienced a marginal gain in LPR and MPR. The L farms of HPR and MPR have increased the share of institutional loans in their total outstanding loans.

Component Analysis of Loan Availment

Cultivator households have availed of loans for the purpose of farm business (FB), non-farm business (NFB), household needs, a small portion for financial business (FNB) and other needs. The issues that are examined are whether the percentage share of these components in total outstanding loan has altered over a period of time, and whether three farm groups exhibit different pattern in these shares. The empirical evidences on these issues are given in Tables 4.5-4.8.

Table 4.5

Component Analysis of Loan Availment: All India

Component	*% Share of Total Outstanding Loan*							
	S		*M*		*L*		*Overall*	
	II	*I*	*II*	*I*	*II*	*I*	*II*	*I*
FB	35 (+)	28	65 (+)	51	68 (+)	58	46 (+)	34
NFB	16 (-)	18	11 (-)	14	8 (-)	12	14 (-)	19
HHD	36 (-)	40	17 (-)	23	16 (-)	20	29 (-)	33
FNB	1 (-)	2	1	1	0 (-)	8	1 (-)	2
Other	12	12	7 (-)	11	8 (-)	9	10 (-)	12
Total	100	100	100	100	100	100	100	100

Note: FB = Farm Business.
NFB = Non-farm Business.
HHD = Household Durables.
FNB = Financial Business.

Source: Estimated by authors.

As evident from Table 4.5, the percentage share of farm business in total loans availed at all-India level has increased by 12 percentage points between 1991-92 and 2002-03, whereas there has been a fall in the share of all other four components.

In case of all the three farm groups, there has been an increase in the share of loan availed for FB during the second period, and a fall in the share of all other components.

Excepting some minor variations, either in overall and across farms groups, the most dominant purpose of loan availment is for FB activity.

The results on component analysis of loan availment by rural cultivator households are given in Table 4.6:

- Overall, FB has increased its availment from 37 per cent to 50 per cent in total outstanding loan over a period of time, and there has been a fall in the share of all other components.

Table 4.6

Component Analysis of Loan Availment: Rural

Component	*% Share of Total Outstanding Loan*							
	S		*M*		*L*		*Overall*	
	II	*I*	*II*	*I*	*II*	*I*	*II*	*I*
FB	39 (+)	37	69 (+)	58	74 (-)	75	50 (+)	37
NFB	13 (+)	12	9 (+)	6	5 (+)	3	11 (-)	12
HHD	35 (-)	37	16 (-)	24	13 (-)	16	28 (-)	37
FNB	1 (-)	2	0 (-)	1	0	0	1 (-)	2
Other	11 (-)	12	7 (-)	11	8 (+)	6	10 (-)	12
Total	100	100	100	100	100	100	100	100

Source: Estimated by authors.

- Credit availment for FB is the most dominant one, in case of all farm groups, and much more so in case of M and L farms.
- There has been a marginal increase in the share of credit availment for NFB and FB in case of S and M farm groups, implying 'activity' diversifaction for diversifying sources of income.

The results (Table 4.7) on component-wise loan portfolio of urban farm households suggest that:

- During the first period, the share of NFB in total outstanding loans activity is the highest (51%), followed by FB activity (21%) and activities related to acquisition of household durable assets. But on the other hand, during the second period, activities related to acquisition of household durable assets are found to have increased the share over other components.
- In the loan portfolio of S farm households, HHD component claims the highest share, followed by NFB and FB, whereas FB claims the highest share in case of M and L farms, and NFB and HHD put together claims 46 per cent in case of M farms and 50 per cent in case of L farms.

Table 4.7

Component Analysis of Loan Availment: Urban

Component	*% Share in Total Outstanding Loan*							
	S		*M*		*L*		*Overall*	
	II	*I*	*II*	*I*	*II*	*I*	*II*	*I*
FB	14 (+)	10	42 (+)	21	42 (+)	14	24 (+)	21
NFB	28 (-)	30	25 (-)	51	21 (-)	37	27 (-)	51
HHD	43 (-)	48	21 (+)	17	29 (-)	31	36 (+)	17
FNB	1 (-)	2	3 (+)	0	0 (-)	2	2 (+)	0
Other	14 (+)	10	9 (-)	12	8 (-)	15	12	12
Total	100	100	100	100	100	100	100	100

Source: Estimated by authors.

It is evident (Table 4.8) from the results on component analysis of loan portfolio in three regions that:

- In LPR, the highest share (54%) is claimed by FB, followed by HHD and NFB. In MPR, the highest share (49%) is also in favour of FB, followed by HHD and NFB. The same pattern is observed in HPR also. In all the three regions of M and L farms, the highest share is claimed by FB.

Table 4.8

Component Analysis of Loan Availment in Regions

Percentage Share in Total Outstanding Loan

Component	*LPR*								*MPR*								*HPR*							
	S		*M*		*L*		*Overall*		*S*		*M*		*L*		*Overall*		*S*		*M*		*L*		*Overall*	
	II	*I*	*II*	*I*	*II*	*I*	*II*	*I*	*II*	*I*	*II*	*I*	*II*	*I*	*II*	*I*	*II*	*I*	*II*	*I*	*II*	*I*	*II*	*I*
FB	41 (+)	36	66 (+)	51	64 (-)	74	54 (+)	48	41 (-)	43	74 (+)	45	52 (+)	38	49 (+)	44	28 (+)	25	60 (+)	58	64 (-)	77	42 (-)	43
NFB	10 (-)	14	7 (-)	11	1 (-)	2	8 (-)	11	17 (-)	18	14 (-)	35	0	0	17 (-)	26	12 (-)	24	11 (-)	12	28 (+)	0	13 (-)	17
HHD	43 (+)	37	22 (-)	29	21 (-)	24	32 (+)	31	30 (+)	29	8 (-)	10	4 (-)	62	25 (+)	21	45 (+)	42	18 (-)	20	8 (+)	7	33 (+)	30
FNB	0 (-)	4	0 (-)	2	0	0	0 (-)	2	0 (-)	2	0	0	0	0	0	1	1	1	0	0	0	0	0 (-)	1
Other	6 (-)	9	5 (-)	8	14 (+)	1	6 (-)	7	11 (+)	7	4 (-)	9	44 (+)	0	9 (+)	8	14 (+)	8	11 (+)	10	0 (-)	15	12 (+)	9
Total	100	100	100	100	100	100	100	100	100	100	100	100	100	100	100	100	100	100	100	100	100	100	100	100

Source: Estimated by authors.

Debt-Asset Ratio Analysis

This section analyses the overall (total) intensity of debt (liability) in relation to cultivator household asset, and examines the intensity with respect to availment of loan from institutional and non-institutional sources. The analysis is performed with all-India data set, data set of rural and urban households, and of data set from three regions. Further, farm group-wise analysis of intensity is also attempted with only the overall debt/total asset ratio, without drawing the distinction between institutional and non-institutional liability. The ratio analysis from all these angles is presented in Tables 4.9-4.11.

As could be observed from Table 4.9:

- The debt-asset ratio (that is liability per unit of asset) is higher during the second period than the first period, suggesting debt intensity of farm households is found to be higher in case of all-India sample of households and rural households. But in case of urban households, the debt intensity has remained the same as between the first and the second period.

Table 4.9

Debt-Asset Ratio: All India, Rural and Urban

Period/Agency	*Ratio*		
	All India	*Rural*	*Urban*
Period II:			
Institutional	0.02	0.02	0.02
Non-institutional	0.01	0.01	0.01
Total	0.03	0.03	0.03
Period I:			
Institutional	0.01	0.01	0.02
Non-institutional	0.01	0.01	0.01
Total	0.02	0.02	0.03

Note: Ratio = Total Outstanding Loan (Liability)/Total Value of Total Assets.
Ratio = Total Institutional Outstanding Loan/Total Assets.
Ratio = Total Non-institutional Outstanding Loan/Total Assets.

Source: Estimated by authors.

- The institutional debt-intensity in relation to asset of all-India households in average is higher during the second period than the first period. The same pattern is observed in case of rural cultivator households. But in case of urban households, not only the total debt intensity has remained the same during both the periods, but also the debt-intensity ratios as between institutional and non-institutional sources have remained the same.

The results on debt-asset ratios of three regions are given in Table 4.10

- The debt-intensity in relation to total assets is found to be higher in the second period in both LPR and MPR, whereas in HPR the intensity has not changed between two periods of time. In case of cultivator households of LPR, the total debt intensity is twice that of MPR and HPR, implying that the cultivator households of LPR have incurred more debt per unit of asset owned by them. Further the share of both institutional and non-institutional agencies in total debt intensity is the same in all the three regions during the second period. The same inference is almost true in case of the share of institutional and non-institutional agencies during the first period.

Table 4.10

Debt-Asset Ratio: Region-wise Analysis

Period/Agency	*Ratio IN*		
	LPR	*MPR*	*HPR*
Period II:			
Institutional	0.02	0.01	0.01
Non-institutional	0.02	0.01	0.01
Total	0.04	0.02	0.02
Period I:			
Institutional	0.01	0.01	0.01
Non-institutional	0.01	0.00	0.01
Total	0.02	0.01	0.02

Note: Same as the ones in Table 4.9.

Source: Estimated by authors.

The debt-asset ratios (the total outstanding loan/total assets) of three farm groups of all-India sample and rural and urban sample are presented in Table 4.11.

- At all-India level, the debt-intensity ratios are higher for S and M farm groups than for L farm group, during the first period, but during the second period, the ratios are the same for all farm groups.
- In case of rural households, the debt-intensity ratios of S and M farm households are found to be higher during the second period, where as the ratios have remained high and constant during both periods in case of L farms.
- With respect to S and M farm of urban households, the ratios are high as well as constant over a period of time. But in case of L farms of urban households, the debt-intensity ratio has decreased during the second period.

Table 4.11

Farm Group-wise Debt-Asset Ratio

Sample/Period	*Farm Group/Ratio*		
	S	*M*	*L*
All India:			
II	0.03	0.03	0.03
I	0.02	0.02	0.01
Rural:			
II	0.03	0.03	0.03
I	0.02	0.02	0.03
Urban:			
II	0.03	0.03	0.02
I	0.03	0.03	0.04

Note: Ratio: Total Outstanding Loan/Total Assets.
Source: Estimated by authors.

Financial Inclusion: Logistic Regression Analysis

The main objective of this section is to answer one major issue: what factors would increase or decrease the probability of financial inclusion of an average farm household of rural, urban and regional (three regions) levels. The major factors considered are education status of the head of the household, age of the head, total non-land farm assets, total non-farm assets, household assets, household size, financial assets and farm size, and region. In case of three regions, LPR is taken as base, and two dummy variables are used for HPR and MPR so as to capture the regional effect on probability of financial inclusion.

The decision rule is: if the odd ratio is positive and greater than one as well as statistically significant, higher probability of increase in financial inclusion is established. If odd ratio is less than one, there is a higher probability of decrease in financial inclusion. The empirical results derived from logistic regression of financial inclusion are given in Tables 4.12–4.15.

As observed in Table 4.12: (a) during the second period, the odd ratios are higher than one for the variables EoH, AoH, value of non-land FBA, household size and farm size, and these are statistically significant. Hence there is a higher probability of financial inclusion of rural cultivator

Table 4.12

Logistic Regression of Financial Inclusion: Rural India

Variables	*Odds Ratio*		*Standard Error*		*t-value*		*P-value*	
	II	*I*	*II*	*I*	*II*	*I*	*II*	*I*
EoH	1.078863	1.157521	0.006315	0.01793	12.97	9.44	***	***
AoH	1.011029	1.004354	0.000896	0.001272	12.38	3.43	***	***
FBA_NL	1.000265	1.000232	0.000025	2.37E-05	10.58	9.79	***	***
NFBA	1.000043	1.000079	2.89E-05	6.61E-05	1.48	1.19		
HHA	0.999985	0.999958	7.42E-06	1.51E-05	-2.01	-2.81	**	***
FA	1.000023	0.998858	0.000135	0.003694	0.17	-0.31		
HHS	1.025379	1.045981	0.004242	0.005913	6.06	7.95	***	***
FS	1.406066	0.89508	0.044491	0.032919	10.77	-3.01	***	***
Log Likelihood	-22601	-10668						
Number of Sample	45054	21700						

Note: EoH = Education of Head.
AoH = Age of Head.
FBA_NL = Non-land Farm Business Assets.
NFBA = Non-Farm Business Assets.
HHA = Household Assets.
FA = Financial Assets.
HHS = Household Size.
FS = Farm Size.
'***' = Level of significance 1%.
'**' = Level of significance 5%.
'*' = Level of significance 10%.

Source: Estimated by authors.

households with an increase in the value of these variables. Further since the odd ratio is less than one for household assets (HHA) and statistically significant, there is a higher probability of financial exclusion with an increase in the value of HHA. (b) During the first period, the inferences are thc same for EoH, AoH, non-land FBA and household size, i.e., there is a higher probability of financial inclusion with an increase in the value of these variables. But the odd ratios for HHA and farm size are less than one. Hence there is a higher probability of decrease in financial inclusion with an increase in the values of HHA and farm size.

With regard to urban farm households (Table 4.13), the odd ratios are greater than one and statistically significant for both the periods in case of EoH and AoH. Hence with an increase in education level of head of the household and the age, there is a higher probability that financial inclusion increases. But during the second period, the odd ratios are greater than one and statistically significant in case of non-land FBA and farm size, and hence the probability of financial inclusion increases with an increase in the value of these variables.

Table 4.13

Logistic Regression of Financial Inclusion: Urban India

Variables	*Odds Ratio*		*Standard Error*		*t-value*		*P-value*	
	II	*I*	*II*	*I*	*II*	*I*	*II*	*I*
EoH	1.111392	1.203106	0.020116	0.038308	5.84	5.81	***	***
AoH	1.013972	1.008107	0.003246	0.003405	4.33	2.39	***	**
FBA_NL	1.000145	1.000113	6.47E-05	7.83E-05	2.24	1.44	**	
NFBA	1.000007	0.999994	1.98E-05	2.57E-05	0.35	-0.24		
HHA	1.000005	0.999994	7.18E-06	1.12E-05	0.76	-0.5		
FA	0.999886	1.002194	0.000124	0.001192	-0.92	1.84		*
HHS	1.015423	1.037744	0.015856	0.016762	0.98	2.29		**
FS	1.404503	1.181727	0.152246	0.130376	3.13	1.51	***	
Log Likelihood	-1667	-1333						
Number of Sample	3241	2335						

Note: Definitions as under Table 4.12.

Source: Estimated by authors.

The regression results (Table 4.14) from the sample of three rural agricultural productivity regions suggest that: (a) during the second period, the odd ratios of EoH, AoH, non-land FBA, household size and farm size are greater than one and statistically significant. Hence there is a higher probability that financial inclusion increases with an increase in the value of these variables. During the same period, the odd ratio for HHA is less than one and statistically significant, implying that as HHA increases there is a higher probability that financial inclusion decreases. (b) During the first period, the odd ratios for EoH, non-land FBA and household size are greater than one and statistically significant. Therefore, with an increase in the value of these variables, the probability of financial inclusion is higher. In case of FS during the same period, the odd ratio is less than one and significant which means that as farm size increases there is a higher probability that financial inclusion decreases. (c) The odd ratio for HPR during the second period is less than one and significant which means that the rural farmers of HPR has a higher probability of financial exclusion compared to that of LPR. But during the first period, the odd ratio for HPR is greater than one and significant. Therefore rural farmers of HPR have higher probability of financial inclusion compared to those of LPR.

Table 4.14

Logistic Regression of Financial Inclusion in Three Regions on Agricultural Productivity of Rural India

Variables	*Odds Ratio*		*Standard Error*		*t-value*		*P-value*	
	II	*I*	*II*	*I*	*II*	*I*	*II*	*I*
EoH	1.082335	1.120102	0.007953	0.025864	10.77	4.91	***	***
AoH	1.010885	0.999508	0.001251	0.00186	8.75	-0.26	***	
FBA_NL	1.000403	1.000312	3.59E-05	3.19E-05	11.22	9.76	***	***
NFBA	1.000007	1.000102	4.58E-05	8.05E-05	0.15	1.27		
HHA	0.999931	0.999905	1.95E-05	3.46E-05	-3.55	-2.75	***	***
FA	0.999519	0.999302	0.000327	0.005597	-1.47	-0.12		
HHS	1.030373	1.055174	0.005178	0.008054	5.95	7.04	***	***
FS	1.286299	0.810289	0.059966	0.053946	5.4	-3.16	***	***
HPR	0.772414	1.36067	0.046266	0.105386	-4.31	3.98	***	***
MPR	0.693339	0.97125	0.028028	0.063163	-9.06	-0.45		
Log Likelihood	-11672	-5121						
Number of Sample	24707	11126						

Note: Definitions as under Table 4.12.
Source: Estimated by authors.

The results on financial inclusion of urban farm households are given in Table 4.15: (a) During both the periods, the probability of financial inclusion increases with an increase in the education level of head and value of non-land FBA. (b) During the first period, the probability of financial inclusion increases with an increase in financial assets and household size. Since the odd ratio for FS is less than one and statistically significant, there is a higher probability that financial inclusion decreases. During the second period, the probability increases with an increase in the age of the head. (c) The probability of financial inclusion of farmers in HPR increases compared to those of LPR during the first period. But during the second period, the urban farmers of MPR have a higher probability of financial exclusion compared to those of LPR.

Table 4.15

Logistic Regression of Financial Inclusion in Three Regions on Agricultural Productivity of Urban India

Variables	*Odds Ratio*		*Standard Error*		*t-value*		*P-value*	
	II	*I*	*II*	*I*	*II*	*I*	*II*	*I*
EoH	1.10885	1.203266	0.022186	0.073291	5.16	3.04	***	***
AoH	1.009565	1.007116	0.005102	0.006371	1.88	1.12	*	
FBA_NL	1.000211	1.000316	9.49E-05	0.000102	2.22	3.11	**	***
NFBA	-	0.99998	5.73E-05	3.79E-05	0.01	-0.53		
HHA	0.999994	0.999988	1.38E-05	1.88E-05	-0.46	-0.64		
FA	1.000321	1.00749	0.000269	0.00271	1.19	2.77		***
HHS	1.036429	1.081575	0.025584	0.027415	1.45	3.09		***
FS	1.158108	0.679736	0.174487	0.149131	0.97	-1.76		*
HPR	1.132282	2.126002	0.190353	0.444213	0.74	3.61		***
MPR	0.63994	1.093258	0.099568	0.233926	-2.87	0.42	***	
Log Likelihood	-766	-502						
Number of Sample	1735	1002						

Note: Definitions as under Tables 4.12 and 4.14.
Source: Estimated by authors.

Debt-Asset Ratio—Explaining Variations: Aggregative Results

One of the issues that needs empirical investigation is why there are variations across farms in debt-asset ratio: total (cumulative) outstanding loan/total asset. It is postulated that variations in ratio could be explained in terms of education status of the head, age of the head, household size, total capital expenditure and farm size. The ordinary least squares (OLS) regression results on this are presented in Tables 4.16–4.19.

It could be observed from the results (Table 4.16) derived with all- India rural household sample that: (a) all the elasticities of debt-asset ratios during both the periods are negative and statistically significant. With an increase in the education status of the head, age, household size, total capital expenditure and farm size, the debt-asset ratios decrease.

Table 4.16

OLS Regression of Debt-Asset Ratio: All-India: Rural

Variables	*Elasticities/ Coefficients*		*Standard Error*		*t-value*		*P-value*	
	II	*I*	*II*	*I*	*II*	*I*	*II*	*I*
EoH	-0.066	-0.193	0.011	0.025	-5.860	-7.87	***	***
AoH	-0.219	-0.070	0.027	0.042	-8.060	-1.66	***	*
HHS	-0.262	-0.283	0.020	0.030	-13.190	-9.43	***	***
EXP	-0.030	-0.084	0.005	0.007	-6.050	-11.17	***	***
FS	-0.325	-0.094	0.039	0.050	-8.270	-1.89	***	*
Constant	-0.028	-0.357	0.108	0.163	-0.260	-2.18		**
Adjusted R Square	0.0126	0.0187						
Number of Sample	45054	21700						

Note: EoH = Log Function of Education of Head.
AoH = Log Function of Age of Head.
HHS = Log Function of Household Size.
EXP = Log Function of Total Capital Expenditure.
FS = Log Function of Farm Size.

Source: Estimated by authors.

Regression results with all-India urban sample are given in Table 4.17. During both the first and second periods, the elasticities have negative values with respect to all the explanatory variables considered in this study. But only household size and farm size are statistically significant during the second period, and education status, household size and total capital expenditure are significant and negative during the first period.

It could be observed from Table 4.18 that: (a) during the second period, debt-asset ratios decrease with an increase in education status, age of the head, household size, capital expenditure and farm size in households of three regions, since all these coefficients are negative during the second period. HPR and MPR are found to have experienced higher coefficients of debt-asset ratio, with LPR as the base. (b) In these regions (rural), during the first period debt-asset ratios decrease with an increase in family size, capital expenditure and farm size. The debt-asset ratio is lower in HPR, compared to that of LPR.

Table 4.17

OLS Regression of Debt-Asset Ratio: All-India: Urban

	Elasticities/ Coefficients		*Standard Error*		*t-value*		*P-value*	
Variables	*II*	*I*	*II*	*I*	*II*	*I*	*II*	*I*
EoH	-0.004	-0.116	0.041	0.059	-0.1	-1.95		*
AoH	-0.177	-0.067	0.125	0.149	-1.41	-0.45		
HHS	-0.162	-0.308	0.065	0.071	-2.5	-4.33	**	***
EXP	-0.023	-0.066	0.015	0.016	-1.5	-3.98		***
FS	-0.333	-0.255	0.111	0.163	-2.99	-1.56	***	
Constant	-0.288	-0.448	0.496	0.566	-0.58	-0.79		
Adjusted R Square	0.006	0.0171						
Number of Sample	3241	2335						

Source: Estimated by authors.

Table 4.18

OLS Regression of Debt-Asset Ratio: Three Regions: Rural

	Elasticities/ Coefficients		*Standard Error*		*t-value*		*P-value*	
Variables	*II*	*I*	*II*	*I*	*II*	*I*	*II*	*I*
EoH	-0.049	-0.018	0.015	0.033	-3.39	-0.55	***	
AoH	-0.229	0.054	0.035	0.062	-6.48	0.88	***	
HHS	-0.277	-0.273	0.022	0.032	-12.53	-8.66	***	***
EXP	-0.040	-0.074	0.008	0.011	-5.26	-6.8	***	***
DHPR	0.265	-0.772	0.042	0.067	6.36	-11.51	***	***
DMPR	0.136	-0.067	0.027	0.042	5.01	-1.61	***	
FS	-0.319	-0.257	0.050	0.051	-6.36	-5	***	***
Constant	-0.023	-0.567	0.138	0.240	-0.17	-2.37		**
Adjusted R Square	0.0181	0.0366						
Number of Sample	24707	11126						

Note: DHPR = Dummy Variable for High Productivity Region.
DMPR = Dummy Variable for Medium Productivity Region. Low Productivity Region is taken as base.

Source: Estimated by authors.

The regression results of debt-asset ratios of sample urban farm households in three regions are given in Table 4.19. (a) During the second period, an increase in household size and in farm size decreases debt-asset ratio, and these elasticity coefficients are statistically significant. The elasticity coefficient for MPR is also negative and statistically significant, with LPR as the base. (b) During the first period, an increase in household size, capital expenditure and farm size decreases debt-asset ratio. The coefficient for MPR is both negative and statistically significant.

Table 4.19

OLS Regression of Debt-Asset Ratio: Three Regions: Urban

	Elasticities/ Coefficients		*Standard Error*		*t-value*		*P-value*	
Variables	*II*	*I*	*II*	*I*	*II*	*I*	*II*	*I*
EoH	0.031	0.005	0.047	0.119	0.65	0.04		
AoH	-0.130	0.168	0.157	0.209	-0.83	0.8		
HHS	-0.167	-0.430	0.081	0.097	-2.05	-4.43	**	***
EXP	-0.032	-0.127	0.020	0.030	-1.58	-4.21		***
DHPR	0.203	0.581	0.126	0.428	1.62	1.36	**	
DMPR	0.094	-0.374	0.118	0.179	0.8	-2.09		**
FS	-0.312	-0.631	0.170	0.154	-1.84	-4.1	*	***
Constant	-0.461	-0.145	0.634	0.143	-0.73	-1.02		
Adjusted R Square	0.0066	0.0529						
Number of Sample	1735	1002						

Source: Estimated by authors.

Debt-Asset Ratio—Explaining Variations: Disaggregative Results

This section analyses determinants of debt-asset ratios with disaggregative results *viz.*, regression results of farm business asset (FBA)–farm business debt (FBD) ratios with all-India and regional rural and urban households. The results are given in Tables 4.20–4.23.

It is observed from Table 4.20 that in case of all-India rural samples, an increase in educational status of the head, in household size, in total capital expenditure and in farm size decreases the FBD-FBA ratios during both the periods. All the negative elasticities are statistically significant during both the periods.

The results (Table 4.21) on elasticities of FBD-FBA ratios of all-India urban households suggest that with an increase in household size, capital expenditure for farm business, and farm size, the FBD-FBA ratios decrease.

Only in case of education status, the ratio increases. These results hold good for both the first and second periods.

Table 4.20

OLS Regression of Farm Business Debt-Asset Ratio: All-India: Rural

Variables	Elasticities/ Coefficients II	Elasticities/ Coefficients I	Standard Error II	Standard Error I	t-value II	t-value I	P-value II	P-value I
EoH	-0.051	-0.036	0.007	0.013	-6.92	-2.7	***	**
AoH	-0.180	-0.035	0.017	0.022	-10.38	-1.57		
HHS	-0.117	-0.109	0.012	0.015	-10.12	-7.32	***	***
FEXP	-0.057	-0.063	0.005	0.006	-12.21	-11.1	***	***
FS	-0.707	-0.287	0.032	0.026	-22.14	-11.04	***	***
Constant	0.536	0.081	0.065	0.086	8.23	0.95		
Adjusted R Square	0.0395	0.0254						
Number of Sample	45054	21700						

Note: FEXP = Log Function of Total Capital Expenditure on Farm Business.
Source: Estimated by authors.

Table 4.21

OLS Regression of Farm Business Debt-Asset Ratio: All-India: Urban

Variables	Elasticities/ Coefficients II	Elasticities/ Coefficients I	Standard Error II	Standard Error I	t-value II	t-value I	P-value II	P-value I
EoH	0.040	0.096	0.022	0.036	1.81	2.69	*	***
AoH	-0.021	-0.003	0.068	0.055	-0.31	-0.05		
HHS	-0.081	-0.086	0.035	0.032	-2.34	-2.73	**	***
FEXP	-0.060	-0.067	0.017	0.017	-3.43	-3.96	***	***
FS	-0.696	-0.475	0.081	0.086	-8.61	-5.54	***	***
Constant	0.018	-0.048	0.279	0.213	0.06	-0.23		
Adjusted R Square	0.0538	0.0457						
Number of Sample	3241	2335						

Source: Estimated by authors.

The results on elasticities of FBD-FBA ratios derived from the sample rural and urban households of three regions are given in Tables 4.22 and 4.23. The results (Table 4.22) on rural households suggest: (a) the FBD-FBA ratios during the second period decrease with an increase in educational status of the head, age, household size, farm size and capital expenditure on farm business. All these elasticity coefficients are negative and statistically significant. The coefficients of FBD-FBA ratios are positive for both HPR and MPR. (b) During the first period, the ratios decrease with an increase in household size, capital expenditure on farm business and farm size. The coefficients are negative for both HPR and MPR and statistically significant. In other words, there are negative regional effects of both HPR and MPR on debt-asset ratios.

Table 4.22

OLS Regression of Farm Business Debt-Asset Ratio: Three Regions: Rural

	Elasticities/ Coefficients		*Standard Error*		*t-value*		*P-value*	
Variables	*II*	*I*	*II*	*I*	*II*	*I*	*II*	*I*
EoH	-0.081	-0.034	0.011	0.022	-7.25	-1.49	***	
AoH	-0.251	-0.008	0.022	0.035	-11.57	-0.23	***	
HHS	-0.104	-0.124	0.016	0.022	-6.63	-5.65	***	***
FEXP	-0.064	-0.062	0.007	0.009	-9.52	-6.97	***	***
DHPR	0.243	-0.184	0.025	0.046	9.67	-4.03	***	**
DMPR	0.094	-0.098	0.016	0.026	5.73	-3.76	***	***
FS	-0.605	-0.334	0.043	0.040	-14.08	-8.27	***	***
Constant	0.717	0.080	0.084	0.132	8.58	0.6	***	
Adjusted R Square	0.0435	0.0287						
Number of Sample	24707	11126						

Source: Estimated by authors.

With respect to results in Table 4.23 on urban farm households: (a) during the second period the FBD-FBA ratios decrease with an increase in household size, capital expenditure on farm business, and farm size. Further, HPR has a positive impact on debt-asset ratio. (b) During the first period, the FBD-FBA ratios decrease with an increase in total capital expenditure on farm business and in farm size. The regression coefficient of FBD-FBA ratio is positive and significant in HPR during the second period, and for MPR it is significant and negative during the first period.

Some Credit-Related Empirical Evidences

This section is a documentation and analysis of results reported in some studies by researchers and by NSSO. Six major questions are examined with empirical evidences reported in these studies. First, what temporal trends have been observed with respect to financial inclusion of cultivator and agriculture labour households? Second, what are the determinants of borrowing from institutional agencies? Third, what are the patterns of interest rate as between institutional and non-institutional agencies in rural India? Fourth, how is the source of credit availment related to asset status of households? Fifth, whether the purpose-wise credit propensity of cultivator and non-cultivator rural households varies? Sixth, how is the pace and pattern of institutional production and investment credit to agriculture sector?

Table 4.23

OLS Regression of Farm Business Debt-Asset Ratio: Three Regions: Urban

	Elasticities/ Coefficients		*Standard Error*		*t-value*		*P-value*	
Variables	*II*	*I*	*II*	*I*	*II*	*I*	*II*	*I*
EoH	0.039	0.052	0.028	0.048	1.38	1.08		
AoH	-0.083	-0.031	0.071	0.091	-1.17	-0.34		
HHS	-0.089	-0.075	0.042	0.053	-2.12	-1.4	**	
FEXP	-0.073	-0.107	0.025	0.039	-2.96	-2.77	***	***
DHPR	0.118	-0.090	0.062	0.075	1.9	-1.2	*	
DMPR	0.086	-0.192	0.055	0.079	1.55	-2.43		**
FS	-0.452	-0.495	0.109	0.142	-4.15	-3.49	***	***
Constant	0.186	0.207	0.286	0.317	0.65	0.65		
Adjusted R Square	0.0402	0.0591						
Number of Sample	1735	1002						

Source: Estimated by authors.

Temporal Analysis of Financial Inclusion and Exclusion of Rural Households

It is recalled from the section: Agency-wise Composition of Household Liability of this chapter that there has been a decline in the share of institutional sources in the total outstanding loans availed by cultivator households as between 1991-92 and 2003. This is much more so in the case of small holders in India. The present exercise is to document and analyse decadal trends in the share of institutional and non-institutional sources over a period of 40 years.

Table 4.24

Decadal Trends in Source of Borrowing by Cultivator Households

Source	*1961-62*	*1971-72*	*1981-82*	*1991-92*	*2002-03*
Institutional	15.8	31.7	63.2	66.3	57.7
Non-institutional	84.2	68.3	36.8	30.7	42.3
Total	100.0	100.0	100.0	100.0	100.0

Source: Adopted from Pallavi (2005).

Two broad era could be observed from the results. First, the period 1961-62 to 1991-92 could be treated as a period of financial inclusion. Second, the period of financial exclusion is found to have been registered during 2002-03. Perhaps, the transition of trends from financial inclusion to exclusion could be attributed to new pattern of growth that has surfaced in India during liquefied petroleum gas (LPG) regime: the growth pattern encompassing accelerated growth of service sector, and decelerated growth of agriculture sector, and almost stagnant growth of industrial sector. This has been much more so during the last one decade or so. In fact, the era of financial exclusion of cultivator households has been much more in states like Punjab, Gujarat and West Bengal.

In Punjab, the share of institutional agencies in outstanding loan of cultivator households had declined from 81.9 per cent in 1991-92 to 47.9 per cent in 2002-03. In Gujarat, the corresponding percentages were 78.5 and 69.5, and in West Bengal these corresponding percentages were 82 and 58. With all the marginal differences in the definition of farm households, the results are conclusive enough to be suggestive of trends towards financial exclusion of cultivator households during recent two decades or so.

Yet, another empirical evidence is related to the share of institutional and non-institutional agencies in total outstanding loan availed by rural cultivator and non-cultivator households, as reported in Part II.A of the National Sample Survey Organisation (NSSO) Report (2007): *Household Indebtedness in India*, based on 2002-03 NSSO All India Debt and Investment Survey (AIDIS) data.

Table 4.25

Distribution of Credit by Credit Agencies: Rural Cultivator and Non-Cultivator Households

	Households/ Agencies	*Percentages*			
		1971	*1981*	*1991*	*2002*
1.	Rural Cultivator Households				
	Institutional	31.7	63.2	66.4	61.1
	Non-institutional	68.3	36.8	33.6	38.9
2.	Rural Non-cultivator Households				
	Institutional	10.8	36.7	55.4	45.9
	Non-institutional	89.2	63.3	44.6	54.1

It is obvious from these empirical evidences that in a span of two decades (1971-1991), there was a rapid transition from financial exclusion to financial inclusion, with an increase in the share of institution credit to rural cultivator households from 31.7 per cent in 1971 to 66.4 per cent in 1991 and in the case of rural non-cultivator households, the increase was from 10.8 per cent to 55.4 per cent, implying higher rate of transition to financial inclusion in case of rural non-cultivator households. But as per the empirical results derived from the AIDIS of 2002-03, there has been a marginal fall in the share of institutional credit to rural cultivator households, and a considerable fall in the share to rural non-cultivator households. In brief, the era of financial inclusion is found to have ended in 1991, and the era of financial exclusion started since then for rural households.

In the context of the issues discussed in this section, an examination of debt profile of agricultural labourers is in order (Pallavi, 2005). Based on the data reported by the Rural Labour Enquiry Committee (1999-2000), the following evidences on the debt profile of landless rural labourers could be examined:

Table 4.26

Percentage Distribution of Debt Outstanding of Agriculture Labour Households: 1993-94 and 1999-2000

	Sources of Debt	*1993-94*	*1999-2000*
1.	Institutional Sources	35.9	31.0
2.	Non-institutional Sources	64.1	69.0
	Total	100.0	100.0

It is obvious from these evidences that there has been a trend towards fall in the share of institutional credit to agriculture labour households, and rise in the share of non-institutional agencies. It is reported (Pallavi, 2005) that the same trend has been observed at the level of states also in India.

Evidences on Determinants of Access to Institutional Credit

Keeping the premise that financial services would form one of the key requirements for income growth, the Reserve Bank of India (RBI) has been promoting financial inclusion through a campaign aimed at providing 'unbanked' households with bank account. In spite of these efforts under social banking for banking the unbanked, a large segment of the society still remains outside the orbit of formal financial system (Thorat, 2006). If the campaign of supply-driven social banking is to achieve the expected outcome of banking the unbanked, an understanding of factors governing demand side becomes crucial (Dev, 2006; Rajalaxmi *et al.*, 2010). It is recalled that the present study (section Financial Inclusions: Logistic Regression Analysis) has performed the empirical exercise, using logistic regression to find out the possible impact of education and age of the head of the household, FBA non-farm business assets (NFBA), household assets, financial assets, farm size and agricultural productivity status of three broad group of regions in the country on financial inclusion. The study by Rajalaxmi *et al.* (2010), using the 2002-03 AIDIS data of NSSO, has considered the impact of micro variable, such as occupation type of household, its social group, its religion, asset ownership, expenditure profile and so on, on access of the household to institutional finance. The major results of this study (as detailed below) are placed in the perspective of the present study.

- Financially excluded households vary from state to state in terms of occupation, land ownership and other socioeconomic characteristics.
- There is lack of strong relationship between access to institutional source of income (supply side) and the percentage of households with debt outstanding (demand side) with formal sources at the state level. In fact, the states with highest outstanding debt tend to have only middle level of access to institutional finance.
- Ownership of land has positive, statistically significant and economically large effect on likelihood of receiving institutional finance.
- In general, scheduled caste (SC), scheduled tribe (ST) and other backward class (OBC) households are less likely be able to access institutional finance.
- In some states, religion has no consistent effect on getting institutional finance, but in some other states, religion has impact on getting institutional finance.

- Monthly per capita expenditure (MPCE) by households has little real impact on the likelihood of receiving institutional loans.

These findings would provide three useful messages in the drive towards financial inclusion. First, financial inclusion has to go much beyond a drive of opening no-frill bank account. Second, a thorough understanding of the characteristics of beneficiary groups is needed for targeting these groups for financial inclusion. Third, the issue of financial inclusion should go beyond getting institutional finance. It should encompass saving pattern, insurance remittances, and participation in money and capital market of the country so as to bring the financially excluded households into the fold of broad-based financial services (Government of India, 2008).

Formation of Interest Rate in Informal Credit Markets

In a major study based on AIDIS data generated by NSSO for the year 2002-03, Manojit and Rajeev (2010) have attempted to answer the question what factors determine interest rate in informal credit market. As a backdrop, some of the results reported in NSSO studies (2007) on cost of debt-terms of interest rate in rural India could be examined.

- Simple interest is the most prevailing mode of terms of debt, and the share of cash debt contracted at simple rate of interest is found to be about 69 per cent in rural areas.
- About 21 per cent of cash debt in rural areas has been contracted at compound interest rate.
- About 2-3 per cent of debt has been advanced at concessional rate, and about 8-10 per cent has been at free of interest rate.
- A high percentage of loans even from institutional agencies has been in the interest range of 12-20 per cent. Obviously, the incidence of interest on loans from non-institutional agencies has been quite high.

As indicated earlier in this section, the major focus of the study by Manojit and Rajeev (2010) is on determinants of interest rate in informal market of developed *vis-à-vis* less developed regions. Following are the major findings and interferences from this study:

- The households in less developed regions are found to have experienced higher rates of interest than those in developed regions. High interest rate is mainly an outcome of monopoly power in the hands of moneylenders in less developed regions, where there are fewer lenders with perfect information about borrowers. But in developed regions, there are more number of moneylenders in the market, and they do not possess full information about borrowers. Further, the borrowers have more option of availing loans from multiple sources, leading to the formation of lower interest rate.
- Education level of borrowers and better economic status are two most important variables that lead to low interest rate in informal credit market. Hence richer rural households have the advantage of availing loans not only from formal credit institutions, but also from informal markets on better terms and conditions.

Credit Availment and Asset Holding

The issue that is examined in this section is: how are asset holding size, source of borrowing and purpose of borrowing associated?

This issue is examined using results reported by NSSO (2007) under sections 8 and 11.C of the reports.

- The share of the households with the lowest asset status in total credit provided by institutional agencies is found to be 0.5 per cent, whereas the share of richest asset stock holding households is 41 per cent.
- Non-institutional agencies have advanced about 14 per cent of the loans to the asset richest classes, whereas 37 per cent of total loans to the poorest asset group.

These results imply two major implications for credit policy to farm households. First, lower the value of asset stock, higher the incidence of indebtedness with non-institutional sources, and higher the value of asset stock, lower the value of incidence of indebtedness. This is to say that the households with lower asset stock depend more on non-institutional sources. However, the richer asset groups have more access to institutional sources. Second, this result also reveals the security-oriented nature of institutional borrowing, rather than the 'social' needs of the household, questioning the very definitions of social banking and financial inclusion.

- About 76 per cent of loan availed by the asset poorest households is found to be for household expenditure, 10 per cent for farm business and 14 per cent for non-farm business. On the other hand, in case of asset richest classes, 59 per cent of cash loan has been for farm business, 13 per cent for non-farm business and 28 per cent for household expenditure.
- In case of the lowest rural asset class, the debt-asset ratio is high to the extent that the value of debt is greater than the value of asset.

Yet, another set of results that should be examined is the pattern of purpose of household loan in rural areas as between cultivator and non-cultivator households.

Purpose	*Cultivator Households*	*Non-cultivator Households*
	Percentages	
Farm business	52.5	9.3
Non-farm business	9.4	19.0
Household needs	27.7	55.0
Others	10.4	16.7
Total	100.0	100.0

Two results could be extracted from the analysis:

- In case of cultivator households, about 53 per cent of loan availed is found to be for farm business, and about 28 per cent for household needs. On the other hand, in case of non-cultivator households, the percentage share of total loan is 55 per cent for meeting household needs, and about 9 per cent for farm business.
- The non-cultivator rural households are found to have borrowed about 19 per cent for non-farm business, and in the case of cultivator households, the percentage share for non-farm business in total loan availed is 9 per cent.

These results validate the widely held impression that the major purpose of loan availed by cultivator households is for farm business activities, and in case of non-cultivator households the dominant purpose is found to be for household needs, followed by non-farm business activities.

Flow of Institutional Credit to Agriculture Sector

It is recalled from earlier sections that the percentage flow of institutional credit to cultivator households has decreased in total credit availed by these households, and that of non-institutional sources has increased during recent years. However it does not mean that the absolute volume of institutional credit to agriculture sector has decreased. Hence, a study of broad trends in the absolute volume of institutional credit to this sector is in order, and an analysis of percentage change in this volume from year to year is also crucial. The analysis on this issue is performed with data for a period of 15 years (1995-96 to 2009-10). One limitation of this analysis is that only trends in nominal values of volume of credit flow are analysed without using the deflator to reduce these nominal values to real values. The results are shown in Table 4.27.

Table 4.27

Trends in Flow of Institutional Credit to Agriculture Sector: 1995-96 to 2009-10

Year	*Type of Credit*					
	Production Credit (₹ crore)	*% Change*	*Investment Credit (₹ crore)*	*% Share*	*Total Credit (₹ crore)*	*% Change*
1995-96	14525	-	7507	-	22032	-
1996-97	16998	17.0	9413	25.4	26411	19.9
1997-98	20640	21.4	11316	20.2	31956	21.0
1998-99	23903	15.8	12957	14.5	36860	15.4
1999-2000	28965	21.2	17303	33.6	46268	25.5
2000-01	33314	15.0	19513	12.8	52827	14.2
2001-02	40509	21.6	21536	10.4	62045	17.5
2002-03	45586	12.5	23974	11.3	69560	12.1
2003-04	54977	20.6	32004	33.5	86981	25.1
2004-05	76062	38.4	49247	53.9	125309	44.1
2005-06	105350	38.5	75136	52.6	180486	44.0
2006-07	138455	31.4	90945	21.0	229400	27.1
2007-08	181393	31.0	73265	-19.0	254658	11.0
2008-09	210461	16.0	91447	24.8	301908	18.6
2009-10	276656	31.5	107858	18.0	384514	27.4

Note: Production Credit = Short-term credit.
Investment Credit = Medium and long term credit.

Source: Various issues of Annual Reports of National Bank for Agriculture and Rural Development (NABARD).

- In a span of 15 years, there has been a 19-fold increase in the flow of production credit to agriculture sector (agriculture and allied sectors) from institutional agencies *viz.*, cooperatives, regional rural banks and commercial banks, and about 14-fold increase in the flow of investment credit. In total, the flow of institutional credit to the sector has been about 17-fold in nominal terms as discussed earlier.
- Barring a few cases, there has been a year to year increase in the percentage change in the flow of production credit to the sector, and year to year fluctuations in the percentage change in flow of institutional investment credit are observed during the last 15 years. Excepting one year of negative percentage change, the range of change has been between 10.4 per cent during 2001-02 compared to 2000-01, and 53.9 per cent during 2004-05 compared to 2003-04.
- In case of total flow of institutional credit to agriculture sector, the percentage change in flow on year to year basis has been positive, with a range of 11 per cent during 2007-08 (compared to 2006-07) to 44.1 per cent during 2004-05 (compared to 2003-04).

The next question which can be examined is what have been the trends in percentage change in the share of production and investment credit to total credit flow. As could be observed from Table 4.28.

- The percentage share of production credit in total institutional credit flow has been a little less than two-third over a periods of time, barring a few cases. In case of investment credit, the trend in percentage share has been a little more than one-third. Excepting a few cases, the percentage shares have remained stable at about two-third and one-third in case of production and investment credit respectively.

Table 4.28

Trends in Percentage Share of Production and Investment Credit in Total Institutional Credit to Agriculture Sector

Year	*Percentage Share*		
	Production Credit	*Investment Credit*	*Total Credit*
1995-96	65.9	34.1	100.0
1996-97	64.4	35.6	100.0
1997-98	64.6	35.4	100.0
1998-99	64.9	35.1	100.0
1999-2000	63.0	37.0	100.0
2000-01	63.1	36.9	100.0
2001-02	65.3	34.7	100.0
2002-03	65.5	34.5	100.0
2003-04	63.2	36.8	100.0
2004-05	60.1	39.9	100.0
2005-06	58.4	41.6	100.0
2006-07	60.0	40.0	100.0
2007-08	71.2	28.8	100.0
2008-09	69.7	30.3	100.0
2009-10	72.0	28.0	100.0

Note: Percentages computed from data in Table 4.27.

Source: Estimated by authors.

Yet another important issue is that have been the trends in allocation of investment credit among sub-sectors/purposes of agriculture sector. As could be observed from Table 4.29.

- Farm mechanisation, high-tech agriculture, animal husbandry and minor irrigation are found to be major areas of institutional investment credit flow.
- Land development, plantation and horticulture, and fishery are found to be next areas of priority for investment credit flow.

In case of S farms (Figure 4.1):

- There has been a decrease in the percentage share of FBAs in total assets with an increase in credit availment during both the periods. Further, the rate of decline in percentage share of FBAs is found to be much more during the second period *viz.*, 2002-03.
- With respect to non-farm business assets, a general trend of increase in the share of NFBAs in total assets with an increase in credit availment over a period of time is observed.
- With an increase in loan availment per unit of cultivated area, there has been an increase at a decreasing rate in the percentage share of household assets during both the periods of the study.
- An increase in credit availment during both the periods has had no impact on the share of financial assets.

In brief, during the year 2002-03, there has been a decrease in the share of FBAs, an increase in the share of NFBAs as well as household assets with an increase in the credit availment, but no change in the share of financial assets.

In case of M farms (Figure 4.2):

- In general, descending ordering of percentage share of each four groups of assets compared to credit availment starts with percentage share of FBAs followed by that of household assets, then by that of NFBAs, and ends with percentage share of financial assets.
- The percentage share of all groups of assets follows a weak ordering in 1991-92. During this period, the percentage share of FBAs has been decreasing at a decreasing rate, and percentage share of household assets has been increasing at an increasing rate, and hence surpasses the percentage share of FBAs. This behavioural pattern defines weak ordering for the year 1991-92.
- During 2002-03, the percentage share of FBAs has a slight downward 'U' shape, whereas the percentage share of household assets has got a 'U shape. Hence, with an increase in credit availment, there has been a shift from FBAs to household assets.

- The percentage share of both NFBAs as well as financial assets has remained almost constant over different amounts of credit availment.

Figure 4.2

Credit Availment and Share of Four Groups of Assets: Rural Medium Farms

(a) 1991-92

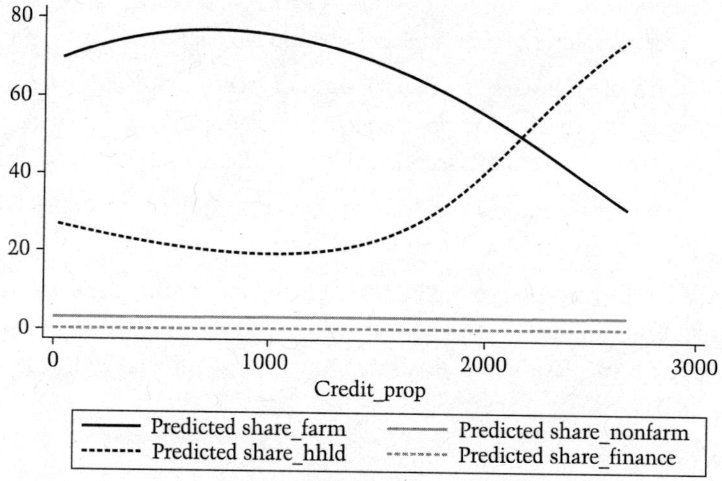

(b) 2002-03

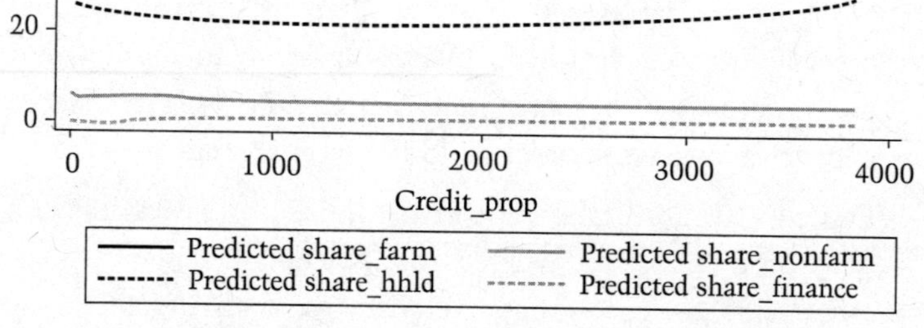

Source: Estimated by authors.

Figure 4.3

Credit Availment and Share of Four Groups of Assets: Rural Large Farms

(a) 1991-92

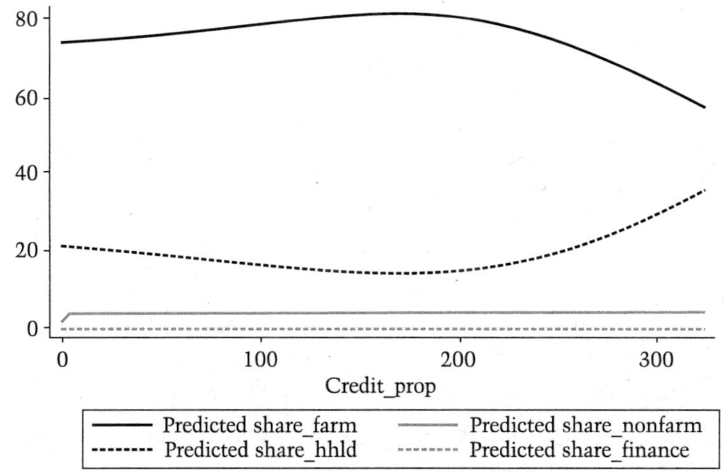

(b) 2002-03

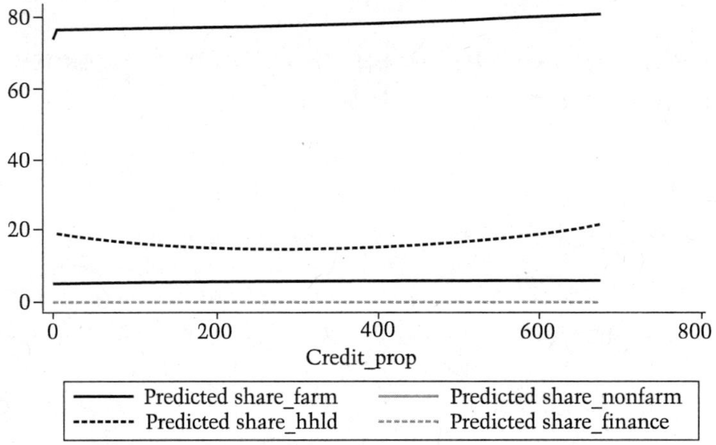

Source: Estimated by authors.

In case of L farms (Figure 4.3):

- The share of FBAs and household assets take highest order compared to other two groups of assets.
- The percentage share of farm business assets increases across propensity to avail credit, but decreases later on for 1991-92, unlike 2002-03.
- The percentage share of household assets is found to be decreasing initially first with an increase in loan per unit of cultivated area, but afterwards there has been an increase in percentage share of household assets during both the periods.
- The percentage share of both financial assets and NFBAs has been not merely low during both the periods, but also has remained constant over credit availment propensity.
- In 1991-92, there has been a shift of share initially to FBAs from household assets, but later this shifting pattern is reversed. In 2002-03, the shift is also from household assets to farm business assets, but during later the percentage share of household assets increases among with farm business assets. This later pattern is found to be due to marginal shift from non-farm business assets.

Farm Business Capital Asset Build-up and Availment of Institutional and Non-institutional Loan

Approach 1: The pictorical presentation of the relationship among FBA, institutional and non-institutional loans per household in rural India is analysed for three groups of farms.

In case of S farms:

- Figure 4.4 depicts a comparative analysis among FBA, IL and NIL variables across two points of time. FBA decreases over the years, but both IL and NIL increases during the years. The increment in NIL is more than IL. Hence, S farms are found to have experienced more financial exclusion.

Figure 4.4

FBA, and Institutional and Non-Institutional Loan per Household, Small Farms of Rural India

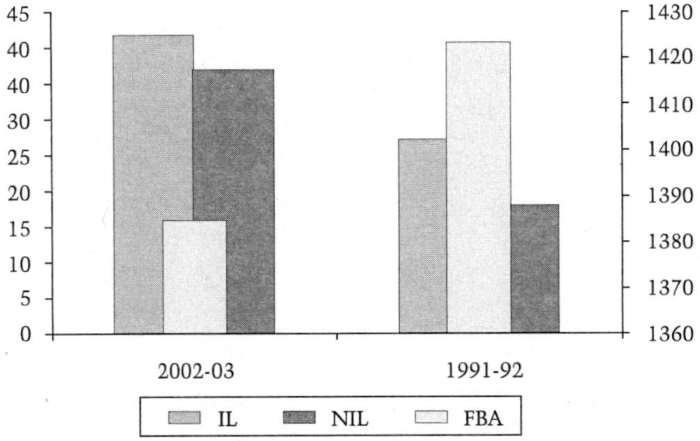

Note: 1. FBA = Total farm business assets per household.
2. IL = Total institutional outstanding loan (in ₹) per household.
3. NIL = Total non-institutional outstanding loan (in ₹) per Household.
4. Left side X-axis is for IL and NIL (average amount of loan).
5. Right side Y-axis is for FBA. These are average amount of FBA.

Source: Estimated by authors.

Figure 4.5

FBA, and Institutional and Non-Institutional Loan per Household, Medium Farms of Rural India

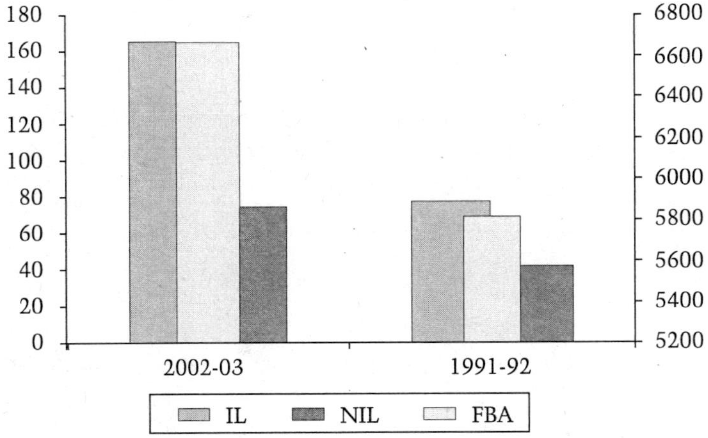

Note: As under Figure 4.4.

Source: Estimated by authors.

In case of M farms:

- FBA increases over the years in M farms, unlike in S farms. Both IL and NIL also increase along the years, but the increment in IL is more than NIL. It can be inferred that M farms have experienced more financial inclusion over time.

Figure 4.6

FBA, and Institutional and Non-Institutional Loan per Household, Medium Farms of Rural India

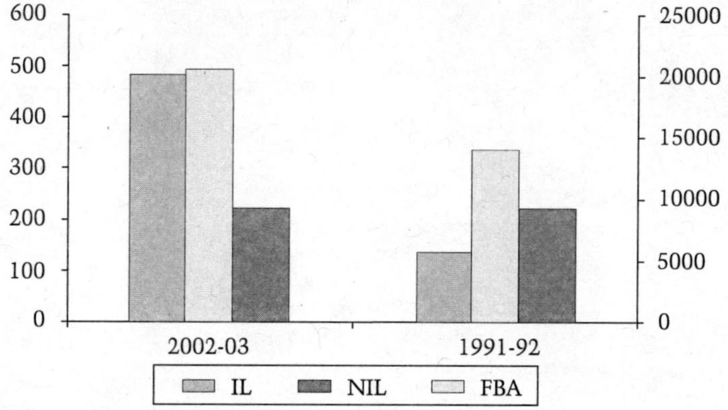

Note: As under Figure 4.4.
Source: Estimated by authors.

In case of L farms:

- In case of L farms, there has been an increase in FBA, IL and NIL over time. The increment in IL is much more than NIL. This is to deduce that L farms like M farms, but unlike S farms, have experienced more financial inclusion over a period of time.

Approach 2: Yet another approach to examine the relationship among FBA, IL and NIL in case of all three farm groups is to use line graphic approach. Figures 4.7-4.11 represent various versions of these relationships.

- As black line slopes from S farms to M farms, FBA increases along with IL. As movement is from M to L farms, on black line FBA still increases with an increase in IL, but intensity of increment decreases. Therefore, when FBA increases, IL also increases along black line, but the intensity is lower later when move is from M to L farms.

- As gray dashed line moves from S to M farms, FBA increases along with NIL, but intensity decreases. Therefore, when FBA increases, NIL also increases but intensity of increment decreases over farm size groups.

Figure 4.7

FBA, IL and NIL Relations in Rural India: 1991-92 (Period I)

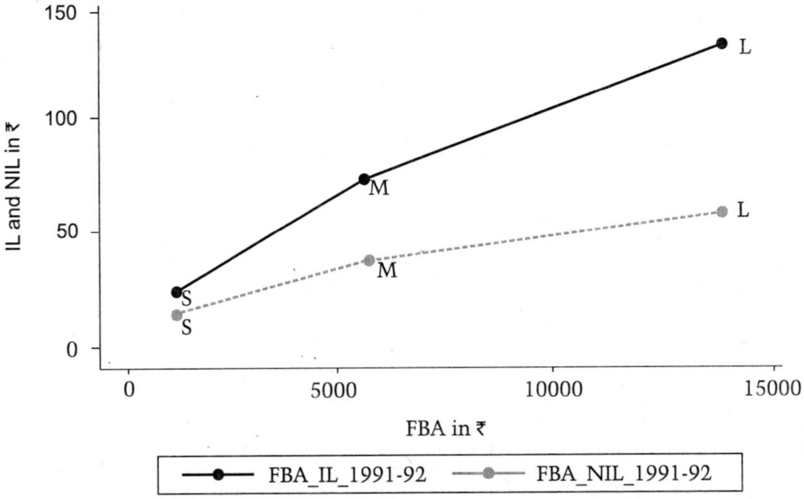

Note: X axis = FBA in ₹.
Y axis = IL and NIL in ₹.
Black line = FBA and IL Relations under three farm groups: S, M and L.
Grey dashed line = FBA and NIL Relations under three farm groups.

Source: Estimated by authors.

A comparison between the movement along black and grey dashed lines suggests that as FBA increases, both IL and NIL increases, but the intensity of increment for IL is higher than for NIL during 1991-92.

- During 2002-03, as black line moves from S to L farms through M farms, FBA increases along with IL and intensity remains the same.
- As grey line moves from S farms to M farms, FBA increases along with NIL.
- As grey line moves from M to L farm groups, FBA increases along with NIL, and the intensity is higher than earlier.

Figure 4.8

FBA, IL and NIL Relations in Rural India: 2002-03 (Period 2)

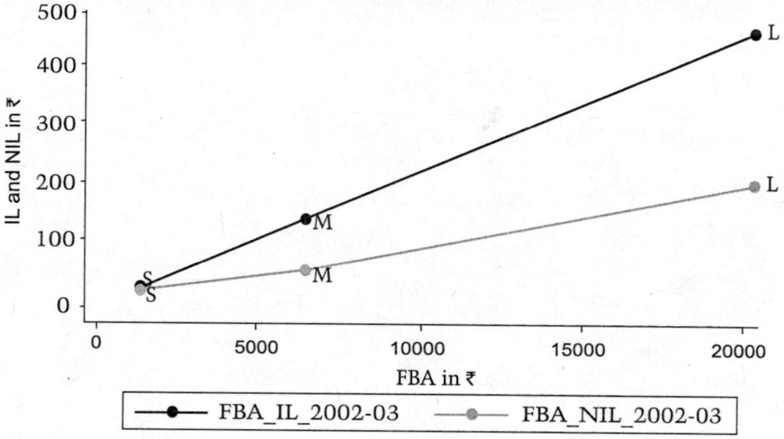

Note: Definitions as under Figure 4.7.
Source: Estimated by authors.

Therefore when FBA increases, NIL also increases and the intensity of increment increases over farm size groups. A comparison of both the black and grey lines reveals that as FBA increases, both IL and NIL increases, but the intensity of increment for NIL is higher than for IL.

Figure 4.9

FBA and IL Relations in Rural India: 1991-92 and 2002-03

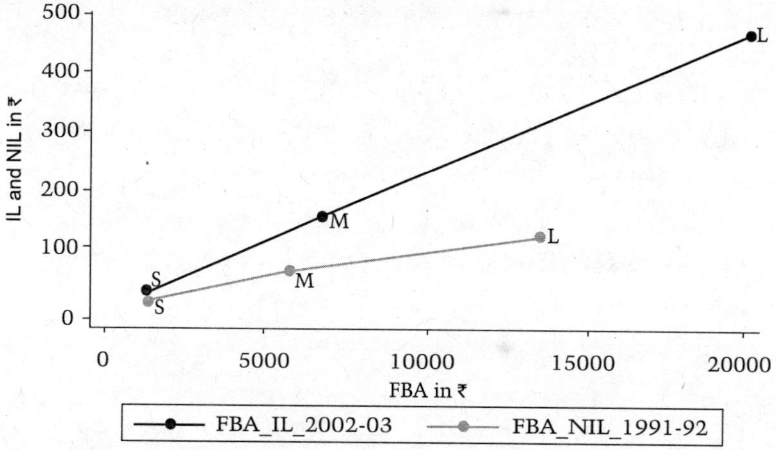

Note: Definitions as under Figure 4.7.
Source: Estimated by authors.

- As black line moves from S to L farms through M farms, FBA increases along with IL at the same intensity.
- As grey line moves from S to M farms, FBA increases along with IL. As grey line lives from M to L farms, FBA increases along with IL, but the intensity of move is higher than earlier. Therefore, when FBA increases, IL also increases and the intensity of increment increases over farm size groups.

A comparison of black and grey lines suggests that as FBA increases, IL also increases for both the time periods but the intensity of increment for IL is higher for the year 2002-03 than for 1991-92.

Figure 4.10

FBA and NIL Relations in Rural India: 1991-92 and 2002-03

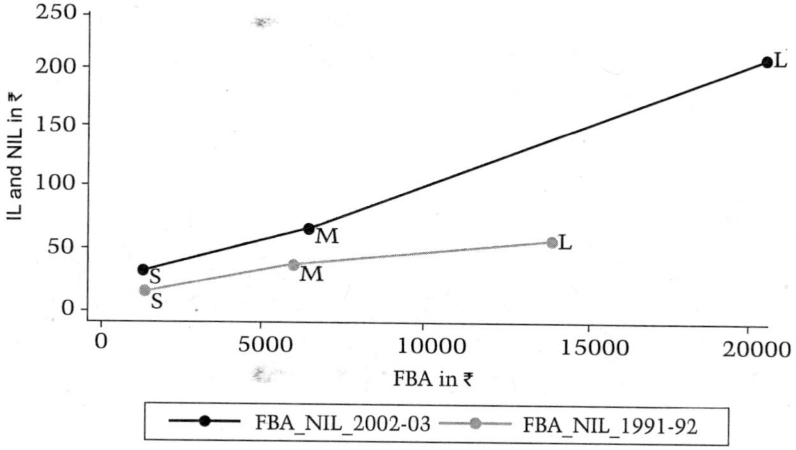

Note: Definitions as under Figure 4.7.
Source: Estimated by authors.

- As black line goes from S to M farms, FBA increases along with NIL. As it goes from M to L farms, FBA increases along with NIL and the intensity is lower than earlier. Therefore, when FBA increases, NIL increases and intensity of increment increases over farms.
- As grey line slopes up from S to M farms, FBA increases along with NIL. As it slopes up from M to L farms, FBA increases along with NIL, but the intensity is lower than it is in case of move from

S to M farms. Therefore, when FBA increases NIL also increases, but the intensity of increment decreases over farms.

Comparison of both black and grey lines suggests that as FBA increases, NIL increases for both the periods of time, but the intensity of increment for NIL is higher for the year 2002-03 than for 1991-92.

Figure 4.11

FBA, IL and NIL Relations in Rural India: 2002-03 and 1991-92

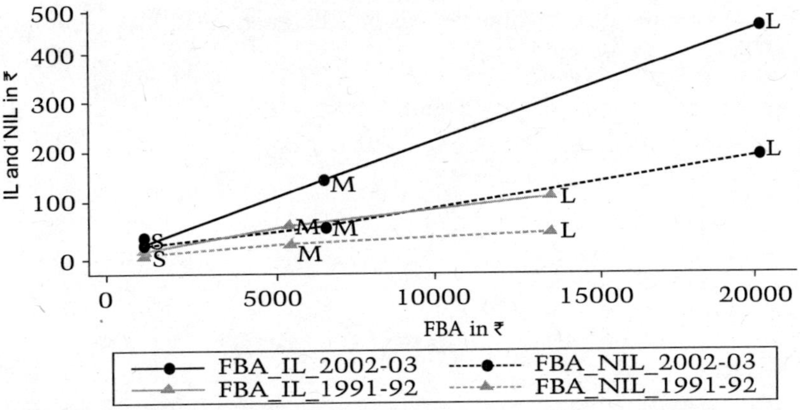

Note: Definitions as under Figure 4.7.

Source: Estimated by authors.

As discussed under Figures 4.7–4.10:

- During the year 1991-92, with an increase in FBA, both IL and NIL increases, but the intensity of increment for IL is higher than for NIL (Figure 4.7). But during the year 2002-03, as FBA increases both IL and NIL increases, but the intensity of increment for NIL is higher than for IL (Figure 4.8).
- As FBA increases, IL also increases for both the time periods, but the intensity of increment for IL is higher for the year 2002-03 than for 1991-92 (Figure 4.9).
- With an increase in FBA, NIL increases for both the periods, but the intensity of increment for the year 2002-03 is higher than for 1991-92 (Figure 4.10).

5 Capital Expenditure

Structure, Growth and Determinants

In this chapter, the focus is on analysing the structure and growth of capital expenditure incurred by farm households at aggregative as well as disaggregative levels. Further, capital asset acquisition and depletion analysis is also performed with results from both aggregative and disaggregative data sets. Added to this is the analysis of capital expenditure debt relations and factors explaining variations in capital expenditure.

Structure of Expenditure on Farm Capital and Non-farm Capital Assets

Capital Expenditure Structure on Farm Business Capital

Tables 5.1-5.8 provide empirical results on structure of capital expenditure on farm assets and non-farm capital assets as well the compositional shifts between 1991-92 (Period I, P.I) and 2002-03 (Period II, P.II). Obviously, these capital expenditures are incurred to build up productive capacity of farm and non-farm activities.

At all-India level (Table 5.1), capital expenditure on farm machinery (FM) is the highest (32%), followed by land development (LD, 23%), animal capital (AC, 21%) and irrigation capital (IC, 18%). Between the two periods, there was an increase in the percentage share of AC and IC, and a decrease in the percentage share of LD and FM.

- In small (S) category of farms, the highest share is found to be on AC, and in other two category of farms the highest share is found to be on FM.
- Over a period of time, S category has experienced an increase in the share of AC and FM, and decrease in the share of LD group

of capital expenditure. With respect to medium (M) category of farms, there has been a gain in the expenditure share of IC and fall in the share of LD. The large (L) farms have registered an increase in the share of LD, IC and AC, but a considerable fall (from 53% to 33%) in the share of FM.

Table 5.1

Composition of Farm Capital Expenditure and Temporal Shifts: All India

Percentage Share

Components	*S*		*M*		*L*		*Overall*	
	II	*I*	*II*	*I*	*II*	*I*	*II*	*I*
LD	27 (-)	34	18 (-)	24	24 (+)	20	23 (-)	27
IC	14 (-)	15	22 (+)	17	23 (+)	18	18 (+)	16
AC	28 (+)	24	13 (-)	14	10 (+)	6	21 (+)	16
FM	24 (+)	20	41	41	33 (-)	53	32 (-)	36
O	7 (+)	6	5 (+)	3	9 (+)	3	6 (+)	4
Total	100	100	100	100	100	100	100	100

Note: LD = Farm Capital Expenditure on Land Development.
IC = Farm Capital Expenditure on Irrigation Capital.
AC = Farm Capital Expenditure on Animal Capital.
FM = Farm Capital Expenditure on Farm Machinery.
O = Farm Capital Expenditure on Others.

Source: Estimated by authors.

With respect to capital expenditure, the rural cultivator households (Table 5.2) exhibit the same structure as in the case of all-India pattern, i.e., highest share of FM followed by LD, AC and IC. Over a period of time, IC and AC have increased their share whereas LD has experienced decrease in its share.

- Farm-group-wise, the pattern as well as inter temporal 'expenditure dominance' is almost the same as that of all-India results.

Table 5.2

Compositions of Farm Capital Expenditure and Temporal Shifts: All India—Rural

Percentage Share

Components	*S*		*M*		*L*		*Overall*	
	II	*I*	*II*	*I*	*II*	*I*	*II*	*I*
LD	27 (-)	32	19 (-)	26	22 (+)	20	23 (-)	27
IC	15 (-)	16	23 (+)	17	27 (+)	15	19 (+)	17
AC	27 (+)	24	14 (-)	15	11 (+)	7	21 (+)	17
FM	24 (+)	21	39	39	36 (-)	55	31 (-)	35
O	7 (+)	6	5 (+)	3	3	3	6 (+)	4
Total	100	100	100	100	100	100	100	100

Note: Definitions as under Table 5.1.
Source: Estimated by authors.

The 'share dominance' in urban cultivator households presents some variation (Table 5.3) compared to that in rural cultivator households.

- The highest share of capital expenditure has gone to FM (37%), followed by AC (24%) and LD (22%). The share of IC is only 8 per cent. Over a period of time, AC has gained considerably from capital expenditure whereas LD and FM category of capital expenditure have experienced a fall in their share. IC in total capital expenditure has experienced a loss of 50 percentage points.
- Three farm groups present different capital expenditure pattern. In S group of households, the most dominant capital expenditure is on AC, in M group on FM, and in L group on LD.
- All the three farm groups have experienced a fall in the expenditure share of IC and the highest fall is in case of L group of farms. Further, all the farm groups have experienced an increase in the share of AC, much more so in case of S category of farms (i.e., from 22% to 40%), which is a 'desirable' trend to provide more additional source of income from animal farming, instead of

depending only on crop production. This is much more so in view of small land-holding size.

Table 5.3

Compositions of Farm Capital Expenditure and Temporal Shifts: All India—Urban

Percentage Share

Components	S		M		L		Overall	
	II	I	II	I	II	I	II	I
LD	22 (-)	46	17 (+)	13	35 (+)	21	22 (-)	28
IC	7 (-)	11	11 (-)	15	5 (-)	44	8 (-)	16
AC	40 (+)	22	6 (+)	5	4 (+)	3	24 (+)	12
FM	23 (+)	15	63 (+)	62	21 (-)	29	37 (-)	38
O	8 (+)	6	3 (-)	5	36 (+)	3	9 (+)	5
Total	100	100	100	100	100	100	100	100

Note: Definitions as under Table 5.1.

Source: Estimated by authors.

It could be observed from the results presented in Table 5.4 that:

- In overall, the highest share (30%) of total capital expenditure incurred by farm households of low-productivity region (LPR) is claimed by AC, whereas in medium-productivity region (MPR) the highest share (47%) is claimed by FM, and in high-productivity region (HPR) also, the highest share (31%) goes in favour of FM.
- In case of S farms of these regions, the highest share (40%) is claimed by AC in LPR as well as HPR, whereas FM claims the highest share (ranging between 73% and 33%) by M category of farms in all the three regions. In case of L category of farms, there is a considerable share in total capital expenditure claimed by FM group of capital assets.
- In overall, IC group has increased its share over a period of time both in LPR and HPR. The same trend is observed in case of AC

Table 5.4

Compositions of Farm Capital Expenditure and Temporal Shifts in Three Regions

Percentage Share

Component	LPR								MPR								HPR							
	S		M		L		Overall		S		M		L		Overall		S		M		L		Overall	
	II	I	II	I	II	I	II	I	II	I	II	I	II	I	II	I	II	I	II	I	II	I	II	I
LD	12 (-)	18	20 (+)	19	31 (+)	17	18 (+)	18	19	19	13 (-)	16	11 (+)	8	17	17	20 (-)	25	19 (+)	18	18 (+)	15	19 (-)	20
IC	26 (-)	29	21 (-)	22	3 (-)	15	22 (+)	21	5 (-)	9	4 (-)	5	24 (-)	27	5 (-)	7	24 (+)	23	25 (+)	15	32 (+)	11	25 (+)	17
AC	40 (+)	35	24 (+)	12	17 (–)	5	30 (+)	14	29 (-)	37	7 (-)	18	14 (-)	38	23 (-)	26	33 (+)	20	12 (+)	11	5 (+)	1	19 (+)	13
FM	19 (+)	15	33 (-)	47	28 (-)	63	26 (-)	45	37 (+)	30	73 (+)	59	50 (+)	20	47 (+)	45	16 (-)	26	41 (-)	53	37 (-)	65	31 (-)	45
O	3	3	3 (+)	1	15 (+)	1	4 (+)	1	9 (+)	5	4	4	1 (-)	7	8 (+)	5	8 (+)	5	4 (+)	3	7 (-)	8	6 (+)	5
Total	100	100	100	100	100	100	100	100	100	100	100	100	100	100	100	100	100	100	100	100	100	100	100	100

Note: Definitions as under Table 5.1.

Source: Estimated by authors.

of these two regions. On the other hand, in these two regions, FM group has experienced decline in its share. There has been not much of a change in the share of land development in total capital expenditure of all the three regions.

Capital Expenditure Structure on Non-farm Business Asset (NFBA)

The next in the scheme of examining empirical results is with respect to composition of non-farm capital expenditure as well as its shift over a period of time (Tables 5.5-5.8).

At all-India aggregative level (Table 5.5), transportation equipment (TE) claims the highest share of 55 per cent in total capital expenditure on non-farm capital assets, followed by building (B) for non-farm business and machinery (Ma) for non-farm business with a share of 19 per cent each.

Table 5.5

Compositions of Non-Farm Capital Expenditure and Temporal Shifts: All India

Percentage Share

Components	*S*		*M*		*L*		*Overall*	
	II	*I*	*II*	*I*	*II*	*I*	*II*	*I*
La	3 (-)	11	7	7	0 (-)	3	3 (-)	10
B	19 (+)	12	20 (+)	18	18 (+)	11	19 (+)	14
Ma	16	16	11 (-)	13	63 (+)	39	19 (+)	17
TE	58 (+)	53	58 (+)	54	18 (-)	33	55 (+)	52
O	4 (-)	8	4 (-)	8	0 (-)	15	4 (-)	8
Total	100	100	100	100	100	100	100	100

Note: La = Non-Farm Capital Expenditure on Land
B = Non-Farm Capital Expenditure on Building
Ma = Non-Farm Capital Expenditure on Machinery
TE = Non-Farm Capital Expenditure on Transportation Equipment
O = Other Non-Farm Capital Expenditure items.

Source: Estimated by authors.

Over a period of time, these three assets are found to have increased their share, whereas land for non-farm business (La) has experienced a decline in its share.

In case of S farms, the highest share (58%) is in favour of TE, and the same pattern (58%) is observed in case of M category of farms; but in case of L category of farms the highest share (63%) is claimed by Ma. These trends are the proxies for diversification of off-farm economic activities.

With respect to rural cultivator households (Table 5.6), the same pattern of 'expenditure dominance' is observed in the sense that TE claims the highest expenditure share (55%), followed by Ma (21%) and B (16%). Further, over a period of time all these three group of assets have claimed increase in their share of capital expenditure.

Table 5.6

Compositions of Non-Farm Capital Expenditure and Temporal Shifts: All India—Rural

Percentage Share

Components	*S*		*M*		*L*		*Overall*	
	II	*I*	*II*	*I*	*II*	*I*	*II*	*I*
La	3 (-)	9	10	10	0 (-)	4	4 (-)	9
B	18 (+)	12	15 (-)	22	7 (+)	3	16 (+)	14
Ma	17 (+)	15	13 (-)	18	75 (+)	55	21 (+)	19
TE	58 (+)	54	57 (+)	41	18 (-)	29	55 (+)	48
O	4 (-)	10	4 (-)	9	0 (-)	9	4 (-)	10
Total	100	100	100	100	100	100	100	100

Source: Estimated by authors.

The S farms operated by rural households have not merely exhibited the 'dominant capital expenditure share' with respect to TE, B and Ma category of assets, and the expenditure share of these three assets has increased in case of S farms.

In case of M group of farms, the dominant expenditure share of 57 per cent is observed in case of TE, followed by B and Ma. But, over a period of time, TE has increased its share, and B and Ma have recorded a decline in their share.

Among L farms, the highest share (75%) of non-farm capital expenditure is in favour of Ma group of assets, and it has increased its share by 20 percentage points between the two periods of the study.

It could be observed from Table 5.7, that the dominant capital expenditure share of 56 per cent is in favour of TE in case of urban cultivator households, followed by B group of assets (12%). B asset has also increased its share considerably from 13 to 29 per cent over a period of time.

Table 5.7

Compositions of Non-farm Capital Expenditure and Temporal Shifts: All India—Urban

Percentage Share

Components	*S*		*M*		*L*		*Overall*	
	II	*I*	*II*	*I*	*II*	*I*	*II*	*I*
La	0 (-)	13	0	0	0	0	0 (-)	10
B	25 (+)	12	30 (+)	12	55 (+)	29	29 (+)	13
Ma	14 (-)	17	5 (+)	1	24 (+)	0	12 (-)	14
TE	58 (+)	51	62 (-)	80	21 (-)	42	56 (-)	57
O	4 (-)	6	3 (-)	8	0 (-)	29	3 (-)	7
Total	100	100	100	100	100	100	100	100

Source: Estimated by authors.

The three groups of urban farms represent the same pattern of share dominance of TE, B and Ma. The S farms have experienced the highest percentage gain in share of B asset group followed by TE asset group. On the other hand, the highest percentage gain is in case of B group of assets of M farms, and the highest loss in share is in case of TE. In case of L group of farms, the highest share (55%) is claimed by B asset group, followed by Ma (24%) and TE (21%).

Two interesting observations could also be made with results in Table 5.7. First, all the three farm groups are found to have increased their share

Table 5.8

Compositions of Non-Farm Capital Expenditure and Temporal Shifts in Three Regions

Percentage Share

Components	*LPR*								*MPR*								*HPR*							
	S		*M*		*L*		*Overall*		*S*		*M*		*L*		*Overall*		*S*		*M*		*L*		*Overall*	
	II	*I*	*II*	*I*	*II*	*I*	*II*	*I*	*II*	*I*	*II*	*I*	*II*	*I*	*II*	*I*	*II*	*I*	*II*	*I*	*II*	*I*	*II*	*I*
La	4 (+)	0	0 (-)	9	0 (-)	0	1	1	3 (-)	24	0 (-)	1	0	0	3 (-)	18	0 (-)	3	0 (-)	9	0 (-)	44	0 (-)	6
B	13 (+)	1	23 (+)	10	1 (-)	9	18 (+)	13	17 (-)	29	27 (-)	49	94 (+)	0	21 (-)	34	31 (+)	3	5 (+)	2	0	0	23 (+)	2
Ma	27 (+)	7	4 (-)	29	0 (-)	4	11 (+)	10	19 (+)	9	12 (-)	13	0	0	18 (+)	10	18 (-)	67	12 (-)	20	91 (+)	0	18 (-)	49
TE	51 (-)	77	73 (+)	39	99 (+)	65	68 (-)	71	55 (+)	33	60 (+)	26	5 (-)	100	55 (+)	31	48 (+)	22	81 (+)	64	9 (-)	54	56 (+)	37
O	5 (-)	15	0 (-)	13	0 (-)	22	2 (+)	15	6 (+)	5	1 (-)	11	1 (+)	0	5 (-)	7	3 (-)	6	3 (-)	5	0 (-)	2	3 (-)	5
Total	100	100	100	100	100	100	100	100	100	100	100	100	100	100	100	100	100	100	100	100	100	100	100	100

Source: Estimated by authors.

of capital expenditure in favour B asset group. Second, both M and L groups of farms have decreased the share of TE and increased the share in case of B and Ma group of assets.

Results on composition of non-farm capital expenditure and the compositional shifts over a period of time in the three regions are given in Table 5.8. Three major results are extracted from this table. First, in all the three regions, the most dominant group that has claimed the highest share (between 68% and 55%) in total capital expenditure on non-farm assets is the TE. Second, the percentage claim of TE is the highest (68%) in LPR. The next dominant one is B category of assets in LPR. In MPR also the dominant share (55%) is claimed by TE, followed by B group of assets (21%). The same pattern is observed in HPR. Third, in all the three regions, the S farms have incurred the highest percentage of their capital expenditure on TE. The same pattern of expenditure dominance is observed in M category farms of all the three regions. In case of L farms, the expenditure dominance of TE is found only in LPR.

Growth Rate Analysis of Capital Expenditure

Growth rate analysis of capital expenditure components is performed with results derived from aggregative and disaggregative data sets as shown in Tables 5.9-5.16, and the growth rate analysis is performed with respect to both farm capital and non-farm capital expenditures.

Farm Capital Expenditure Growth Rate Analysis

At all-India level (Table 5.9), cutting across three farm groups and five components of farm capital expenditure growth rate is found to be 2.4 per cent. Barring the growth rate of 'miscellaneous capital expenditure group,' the growth rate of expenditure on AC asset is found to be the highest (4.8%), followed by the next highest of 3.1 per cent expenditure growth on IC. The growth rates of capital expenditure on LD and FM groups of farm business assets are around 1.0 per cent.

With regard to the three farm groups, S and L farms have experienced capital expenditure growth rate of over 6.0 per cent in aggregate, but M group of farms only 0.4 per cent. The growth rates in case of AC and FM are 8.0 per cent each, in S farms, and capital expenditure growth rates of 5.5 per cent and 4.0 per cent are observed respectively in case of LD and IC group of assets. In case of M farms, the moderate expenditure growth rate of 2.9 per cent is found under IC assets. L farms have experienced negative growth rates of capital expenditure under FM, LD, IC and AC group of assets.

Table 5.9

Growth Rates of Farm Capital Expenditure: All India

Components	*S*	*M*	*L*	*Overall*
LD	4.0	-2.1	-5.0	0.9
IC	5.5	2.9	-4.1	3.1
AC	8.0	0.1	-2.6	4.8
FM	8.0	0.4	-10.4	1.2
O	8.0	3.5	3.8	6.0
Total	6.4	0.4	-	2.4

Source: Estimated by authors.

The growth rates of capital expenditure estimated with the sample of only rural cultivator households are shown in Table 5.10.

- The overall growth rate of 2.5 per cent has occurred in case of rural cultivator households. The capital expenditure growth rates are around 4.0 per cent under IC and AC groups of assets, and quite low under LD and FM groups of assets.
- S farms have recorded the highest growth rate (6.9%) of farm capital expenditure, and as high as around 8 per cent in the capital components of FM and AC.

Table 5.10

Growth Rates of Farm Capital Expenditure: All India—Rural

Components	*S*	*M*	*L*	*Overall*
LD	5.2	-2.3	-6.6	1.1
IC	6.0	3.4	-2.4	3.8
AC	7.9	0.1	-2.8	4.5
FM	8.1	0.7	-10.9	1.3
O	8.3	4.4	-5.7	6.0
Total	6.9	0.6	-7.4	2.5

Source: Estimated by authors.

With regard to M farms, the growth rate in aggregate is only 0.6 per cent, with a growth rate of 3.4 per cent under IC assets. The L farms of rural households are found to have registered negative growth rates of capital expenditure in total as well as component wise.

The results on farm capital expenditure growth rates of urban cultivator households (Table 5.11) suggest that:

- Even though expenditure growth rate of 1.0 per cent is observed in total, negative expenditure growth rates are seen in case of M and L groups of farms, with only S farms recording a growth rate of 3.0 per cent in capital expenditure on farm business assets (FBA). With respect to component-wise capital expenditure growth, AC has secured an expenditure growth rate of 7.4 per cent, and IC, LD and FM groups of assets experiencing either low or negative expenditure growth rates in capital expenditure, cutting across farm groups.
- In case of S farms, the growth rates of capital expenditure on AC and FM have been not merely positive, but also quite high.
- Both M and L farms have registered negative growth rates of expenditure under IC, FM and AC, excepting in case of L farms where the growth rate is positive under AC.

Table 5.11

Growth Rates of Farm Capital Expenditure: All India—Urban

Components	*S*	*M*	*L*	*Overall*
LD	-3.5	1.1	4.6	-1.4
IC	-1.6	-3.8	-18.5	-5.3
AC	8.8	-0.5	1.8	7.4
FM	6.7	-1.0	-3.0	0.7
O	5.7	-5.2	24.1	5.9
Total	3.0	-1.2	-0.1	1.0

Source: Estimated by authors.

In total, the growth rates of capital expenditure in all the three regions are found to be positive, with the highest (5.3%) rate recorded in MPR and the lowest (2.1%) rate in LPR (Table 5.12).

- Among the three regions, the S farms of LPR and MPR have registered a capital expenditure growth rate of over 10 per cent. The expenditure growth rates are not merely positive under LD, IC, AC and FM groups of assets in all the three regions, but also as high as around 13 per cent in LPR and MPR.
- The total capital expenditure growth rates in case of M farms are either low or negative in the study regions. Only a high growth rate of 7.8 per cent is found in LPR under AC asset group.
- The L farms are found to have experienced a negative growth rates of total capital expenditure in all the three regions with some exceptions of positive in case expenditure growth rates under AC in LPR and HPR and under IC in HPR.

Table 5.12

Growth Rates of Farm Capital Expenditure in Three Regions

Components	*LPR*				*MPR*				*HPR*			
	S	*M*	*L*	*Overall*	*S*	*M*	*L*	*Overall*	*S*	*M*	*L*	*Overall*
LD	6.5	1.7	-1.9	2.0	11.0	-3.0	-11.1	5.7	3.1	4.0	1.0	3.3
IC	9.1	0.8	-12.5	2.4	5.2	-3.2	-14.6	1.7	5.6	8.6	9.4	7.4
AC	11.5	7.8	3.0	-9.3	8.4	-8.9	-21.5	4.0	10.0	4.3	14.6	7.9
FM	12.4	-1.9	-14.0	-2.8	13.1	0.7	-6.1	5.6	0.8	1.4	-5.6	0.3
O	11.3	15.1	18.7	14.9	18.4	-0.8	-26.3	11.7	9.6	5.3	-1.8	6.1
Total	10.2	1.3	-7.5	2.1	10.9	-1.2	-13.7	5.3	5.4	3.7	-0.8	3.9

Source: Estimated by authors.

Growth Analysis of Non-Farm Capital Expenditure

The results on estimated growth rates of capital expenditure on non-farm assets both in aggregative and disaggregative frame are presented in Tables 5.13-5.16. The major inferences derived from these tables are as detailed below:

- As could be seen from Table 5.13, all the three farm groups at all-India level have experienced a positive growth rate of total non-farm capital expenditure. Obviously growth rates vary across farm groups as well as expenditure components.

Table 5.13

Growth Rates of Non-farm Capital Expenditure: All India

Components	*S*	*M*	*L*	*Overall*
La	-8.5	5.1	-100.00	-4.3
B	9.2	5.3	13.6	8.5
Ma	4.8	3.4	13.0	6.1
TE	5.5	5.4	2.5	5.4
O	-2.4	-1.8	-24.3	-3.1
Total	4.6	4.7	8.1	4.8

Source: Estimated by authors.

- At all-India level, across components, the capital expenditure growth rate is the highest (8.5%) under B group of assets, followed by Ma and TE group of assets.
- Component-wise the growth rates of expenditure on B, Ma and TE groups of non-farm business assets are positive and high in all the three farm groups.

• The growth rates of non-capital expenditure in case of rural cultivator households are given in Table 5.14:

- The total capital expenditure growth rates in overall as well in all three farm groups have been both positive and quite high (as high as 8.0%).
- The S farms have recorded a total capital expenditure growth rate of 8.9 per cent, with the highest rate (13%) being in case of B asset, followed by Ma asset group (10.4%) and TE group (9.7%).
- The M farms are found to have experienced a moderate growth rate of 5.3 per cent, and the highest being in TE (8.5%), and the lowest in B group of asset.
- In case of L farms, the total capital expenditure growth rate has been as high as 8.8 per cent, the highest growth rate (17.5%) recorded in case of B asset and the lowest positive growth rate recorded under TE group.

Table 5.14

Growth Rates of Non-farm Capital Expenditure: All India—Rural

Components	*S*	*M*	*L*	*Overall*
La	-1.2	5.1	-100.0	1.0
B	13.0	1.9	17.5	9.5
Ma	10.4	2.4	12.0	9.1
TE	9.7	8.5	3.9	9.2
O	-0.2	-1.3	-18.0	1.0
Total	8.9	5.3	8.8	8.0

Source: Estimated by authors.

- The results (Table 5.15) on capital expenditure growth rates of NFBA of urban cultivator households suggest that:
  - In overall, there has been a negative growth rate of total capital expenditure, with positive growth rate of 6.4 per cent in B component.
  - The S farms of urban area have experienced a negative growth rate in total as well as component-wise expenditure growth rates, excepting a positive growth rate of 3 per cent in B component.
  - The capital expenditure growth rate of M farms of urban cultivator households has been 3.5 per cent in total, the highest growth rate (15.8%) in case of Ma component and the lowest (1.1%) in case of TE component.
  - The L farms have recorded a growth rate of 6.1 per cent in total, and component-wise the highest being 12.5 per cent in case of B component.

Table 5.15

Growth Rates of Non-farm Capital Expenditure: All India—Urban

Components	*S*	*M*	*L*	*Overall*
La	-100.0	0.0	0.0	-100.0
B	3.0	12.4	12.5	6.4
Ma	-5.7	15.8	0.0	-2.7
TE	-2.5	1.1	-3.0	-1.3
O	-8.0	-3.4	-100.0	-8.3
Total	-3.5	3.5	6.1	-1.3

Source: Estimated by authors.

- Table 5.16 presents results on non-farm capital growth rates of the three regions:
  - Total growth rates in all the three regions have been positive, and it has been over 7.0 per cent in LPR.
  - In LPR, the expenditure growth rate is the highest (27.8%) in case of B component and lowest (6.0%) in case of La component.

    In MPR, the growth rates have been more than 12.0 per cent in Ma and TE components, and it has been the lowest (2.1%) in case of B component.

    In HPR, the highest positive growth rate of a little over 29 per cent is in B component and the lowest positive growth rate of over 9 per cent in case of TE component.
  - The S farms of both MPR and HPR are found to have experienced positive growth rates of non-farm capital expenditure in total, whereas in LPR the growth rate has been negative. In case of both M and L farms of all the regions, the total growth rates have been positive, and it is 22.4 per cent in M farms of LPR.
  - Component-wise, the growth rate of capital expenditure in B component of small farms in LPR is the highest (31.6%), followed by La component (26.3%) and Ma component (11.5%). These farms in MPR have recorded the highest growth rates of 15.5 per cent and 13.0 per cent respectively in case of Ma and TE components. In HPR, La component has experienced the highest growth rate of 33.1 per cent.
  - With regard to component-wise growth rates in Ma farms of MPR, the highest positive growth rate (10.8%) is in TE component, and in HPR the highest growth rate (10.6%) is in B component.
  - In case of L farms of all the regions, component-wise growth rates have been mostly negative or insignificant.

Table 5.16

Growth Rates of Non-farm Capital Expenditure in Three Regions

Components	*LPR*				*MPR*				*HPR*			
	S	*M*	*L*	*Overall*	*S*	*M*	*L*	*Overall*	*S*	*M*	*L*	*Overall*
La	26.3	-100.0	0.0	6.0	-10.2	-100.0	0.0	-10.4	-100.0	-100.0	-100.0	-100.0
B	31.6	32.0	-10.8	27.8	2.8	-2.7	0.0	2.1	33.1	10.6	0.0	29.1
Ma	11.5	1.8	-26.8	8.0	15.5	1.6	0.0	12.1	-6.0	-1.3	0.0	-4.0
TE	-4.7	29.7	11.1	6.8	13.0	10.8	-2.2	12.5	13.9	5.9	-12.1	9.3
O	-10.8	-11.6	-100.0	-12.0	8.4	-20.2	0.0	3.1	0.1	-2.1	-100.0	-0.7
Total	-1.1	22.4	7.0	7.1	7.9	2.6	-	6.8	6.0	3.7	3.4	5.2

Source: Estimated by authors.

Fixed Capital Expenditure Pattern in Cultivator Households

An examination of distribution of fixed capital expenditure by cultivator households during a period of two decades would provide insights into the dynamics of expenditure shifts (Bisaliah, 2010c). It could be observed from Table 5.17 that:

- Agricultural machinery and equipment would form the most important item of capital expenditure, followed by expenditure on wells and other irrigation structures at all-India level. These two items alone would form a little over 70 per cent of the capital expenditure incurred by the household sector in agriculture during the period of 20 years. Almost the same pattern of capital expenditure by cultivator household is also observed at the state level for 1991-92 as per the All-India Debt and Investment Survey (AIDIS) (Gulati and Bathla, 2002).
- Capital expenditure on land improvement and farm houses and animal sheds would account for next expenditure item of priority.
- Capital expenditure on orchards and plantations has increased substantially over a period of time.

Table 5.17

Percentage Distribution of Fixed Capital Expenditure of Cultivator Households: All-India Estimates between 1971-72 and 1991-92

Capital Asset	Percentage Share		
	1971-72	1981-82	1991-92
1. Machinery and Equipment	43.20	51.95	47.80
2. Wells and other irrigation structure	26.80	20.45	24.70
3. Land improvement	16.50	15.25	12.91
4. Farm houses and animal sheds	10.50	4.55	3.67
5. Orchards and plantations	1.80	2.80	5.08
6. Other capital assets	1.20	5.00	6.26
Total	100.00	100.00	100.00

Source: Gulati and Bathla (2002).

Whether the same patterns and priorities of capital expenditure have been registered as per AIDIS of 2002-03 are displayed in Table 5.18. Further, the patterns and priorities of capital expenditure in six states of three agricultural productivity regions are also analysed in the same table.

Table 5.18

Percentage Distribution of Fixed Capital Expenditure on Farm Business Assets in Rural Cultivator Households: India and in the States of Three Agricultural Productivity Regions: 2002-03

Asset	All India	Productivity Regions/States					
		HPR		MPR		LPR	
		TN	Punjab	UP	Bihar	MP	Rajasthan
1. Land improvement	5.9	5.1	2.1	0.4	2.2	9.5	3.7
2. Orchards and plantations	1.5	0.2	0.0	0.0	0.0	0.0	0.0
3. Wells and other irrigation structures	33.1	65.0	30.1	3.2	5.1	31.3	52.3
4. Machinery and equipment	22.1	11.7	23.1	36.8	30.9	17.2	11.3
5. Farm houses and animal sheds	36.0	13.6	44.5	59.2	61.0	41.4	31.4
6. Other capital expenditure	1.4	4.4	0.2	0.4	0.8	0.6	1.3
Total	100.0	100.0	100.0	100.0	100.0	100.0	100.0

Source: Estimated using NSSO unit level data for the year 2002-03.

- At all-India level, the dominant items of capital expenditure are found to be farm houses and animal sheds, wells and other irrigation structures, and machinery and equipment. These alone would account for 91 per cent of total fixed capital expenditure. Land improvement comes next in importance in claiming its share in total expenditure. Compared to share in 1991-92, orchards and plantations have experienced a fall in their share. It is understandable that machinery and equipment (labour substitutes), and wells and other irrigation structures (land substitutes) would continue to be dominant items of capital expenditure for years to come.
- In HPR, Tamil Nadu (TN) represents a case of dominant capital expenditure on assets of wells/other irrigation structure, and Punjab a case of farm houses/animal sheds and wells/other irrigation structure dominance.
- In MPR, Uttar Pradesh (UP) represents a typology house/animal shed, and machinery/implements expenditure dominance. Bihar also represents the same dominance typology.
- In LPR, Madhya Pradesh (MP) represents a pattern of farm house/ animal sheds, and irrigation wells/other irrigation structure dominance in capital expenditure. Rajasthan also falls into the same typology pattern.

One major inference from the analysis of results on distribution of capital expenditure on fixed capital assets is that irrigation, machinery and farm houses/animal sheds continue to establish their 'dominance' at the national as well as regional levels during the last three decades of 1971-2002.

Asset Acquisition and Depletion Analysis

Asset acquisition analysis is performed with available data for the year 1991-92, whereas asset depletion (sale and loss due to natural disaster and so on) analysis is performed with data for both the periods. The analysis is with both aggregative and disaggregative results.

Asset Acquisition Pattern

It could be seen from Table 5.19 that:

- Acquisition of FBA forms the single most important item (over 50%) in total assets acquired by farm households both at all-India and rural levels and the next highest is the household asset (HHA). The percentage share of NFBA in total acquisition is not that high. But in case of urban farm households, the highest share in total asset acquisition is in the case of HHA (34%), followed by FBA (28%) and NFBA (23%). Further, NFBA has claimed a higher share (23%) in case of urban households, than in case of rural households (9%). In other words, there is a greater move of urban farmers towards asset diversification. This is also reinforced by 58 per cent share of FBA in asset acquisition pattern of rural households, whereas it is only 28 per cent in case of urban households.

  The small farmers, at all-India level as well as rural, have revealed their preference for FBA, whereas the preference of urban households for FBA is as low as 17 per cent and their preference is more towards HHA and NFBA. The small farmers in general have acquired more of HHA than NFBA in total asset acquisition.

  The M group of farmers, both at all-India level and rural/urban level, have revealed their preference more for FBA than for other assets. The next in order of preference is HHA.

Table 5.19

Compositions of Acquisition of Assets: All India, Rural, Urban

Percentage Share

	India				*Rural*				*Urban*			
	S	*M*	*L*	*Overall*	*S*	*M*	*L*	*Overall*	*S*	*M*	*L*	*Overall*
	I	*I*	*I*	*I*	*I*	*I*	*I*	*I*	*I*	*I*	*I*	*I*
FBA	41	64	55	52	51	66	58	58	17	54	16	28
NFBA	15	8	12	12	9	7	12	9	29	10	15	23
HHA	37	21	28	30	36	21	28	28	41	16	33	34
FA	7	8	4	7	5	6	2	5	13	19	35	16
Total	100	100	100	100	100	100	100	100	100	100	100	100

Source: Estimated by authors.

In case of large farmers, the rural households have preferred FBA over other assets, whereas urban households have preferred HHA and FA. All the farm groups at all levels have preferred more of HHA than NFBA.

Results on asset acquisition pattern of farm households in three regions are shown in Table 5.20.

- In all the regions, FBA secures the highest share (50%) in total volume of expenditure incurred on asset acquisition, followed by HHA (ranging between 34% and 29%) and NFBA (with a range of 14%-6%).
- The share of FBA in case of S farmers of the three regions would be around 40 per cent of total asset acquired, in case of M group of farmers the share would come to around 60 per cent, and the range of share of FBA is found to be between 61 per cent in HPR, 52 per cent in LPR and 29 per cent in MPR. The next asset preference of all farm groups of all the regions is HHA, with an exception that large farmers of MPR have preferred FA.
- The farmers of HPR, cutting across farm groups, are in favour of accumulating relatively a good percentage of their assets in FA.

Table 5.20

Compositions of Acquisition of Assets in Three Regions

Percentage Share

	LPR				*MPR*				*HPR*			
	S	*M*	*L*	*Overall*	*S*	*M*	*L*	*Overall*	*S*	*M*	*L*	*Overall*
	I	*I*	*I*	*I*	*I*	*I*	*I*	*I*	*I*	*I*	*I*	*I*
FBA	43	64	52	55	48	63	29	53	38	61	64	50
NFBA	13	11	7	10	16	10	33	14	9	4	0	6
HHA	42	25	40	34	35	22	7	29	42	18	25	31
FA	2	0	1	1	1	6	31	4	11	17	10	14
Total	100	100	100	100	100	100	100	100	100	100	100	100

Source: Estimated by authors.

Asset Depletion Pattern

The asset depletion analysis is performed with data for both the periods, unlike the asset acquisition analysis performed with data of the first period of this study. The results are given in Tables 5.21-5.23.

Overall, FBA at all-India level accounts for around 60 per cent in the total value of depleted assets. The next group of asset in accounting for depletion is HHA with a share of 35 per cent.

The depletion share of FBA of M and L group of farmers comes to 67 per cent, whereas it comes to 56 per cent in case of S farms.

Over a period of time, all the farms have experienced a decline in the share of FBA and NFBA, whereas HHA would account for an increase in the share in total depletion of assets (Table 5.21).

Table 5.21

Compositions of Depletion of Assets and Temporal Shifts: All India

Percentage Share

	S		*M*		*L*		*Overall*	
	II	*I*	*II*	*I*	*II*	*I*	*II*	*I*
FBA	56 (-)	64	67 (-)	76	67 (+)	39	60 (-)	66
NFBA	6 (-)	12	1 (-)	5	19 (-)	32	5 (-)	11
HHA	38 (+)	19	32 (+)	16	15 (-)	26	35 (+)	18
FA	0 (-)	6	0 (-)	3	0 (-)	3	0 (-)	5
Total	100	100	100	100	100	100	100	100

Source: Estimated by authors.

It could be seen from Table 5.22 that the same pattern of asset depletion has occurred in rural cultivator households with the 'depletion dominance' of FBA followed by HHA, and in the case of L farms the share of FBA in total value of depletion is as high as 83 per cent.

The share of HHA has increased in total depletion under both S and M farms, whereas in L farms the share has declined.

The share of NFBA has declined over a period of time, in total depletion in case of all the three farm groups.

Table 5.22

Compositions of Depletion of Assets and Temporal Shifts: All India—Rural

Percentage Share

	S		M		L		Overall	
	II	I	II	I	II	I	II	I
FBA	59 (-)	68	69 (+)	80	83 (+)	39	63 (-)	70
NFBA	5 (-)	11	1 (-)	3	0 (-)	31	4 (+)	10
HHA	35 (+)	19	30 (+)	15	17 (-)	27	33 (+)	18
FA	0 (-)	3	0 (-)	3	0 (-)	3	0 (-)	3
Total	100	100	100	100	100	100	100	100

Source: Estimated by authors.

The results (Table 5.23) on asset depletion pattern of urban cultivator households present slightly different picture. In overall, the share of HHA in total assets depleted is 50 per cent, whereas it is 38 per cent in case of FBA, and the share of NFBA in total depletion is found to be 12 per cent.

Table 5.23

Compositions of Depletion of Assets and Temporal Shifts: All India—Urban

Percentage Share

	S		M		L		Overall	
	II	I	II	I	II	I	II	I
FBA	35 (-)	50	53 (+)	39	12 (-)	31	38 (-)	47
NFBA	6 (-)	14	4 (-)	26	81 (+)	53	12 (-)	18
HHA	59 (+)	20	43 (+)	24	8 (-)	15	50 (+)	21
FA	0 (-)	16	0 (-)	10	0	0	0 (-)	14
Total	100	100	100	100	100	100	100	100

Source: Estimated by authors.

Farm-group-wise analysis over a period of time suggests that: (a) there has been a decline in the share of FBA, but an increase in the share of HHA in total value of assets depleted under S farm category. (b) In case of M group of farms, there has been an increase in the share of FBA as well as HHA, and decline in the share of NFBA. (c) There has been a decline in the share of FBA and HHA, and a considerable increase in the share of NFBA under L farm category.

The asset depletion analysis of the three regions is shown in Table 5.24.

In the total value of assets depleted, FBA forms the most important group in all the regions in terms of its share of over 60 per cent. The next in importance in all the three regions is HHA. But FBA has improved its position during the second period, by contributing lower share to the total value of assets depleted in all the regions.

In case of S farms, there has been a decline in the share of FBA to total value of assets depleted in LPR and HPR. The M farms have experienced a fall in the share of FBA in all the regions, but the share of FBA of L farms has increased in both LPR and HPR.

There has been an increase in the share of HHA in total value of assets of S as well as M farms of all the regions, but this trend is true of L farms only in LPR and MPR.

Typologies of Debt-Asset Acquisition and Depletion: Farm-Size-Wise Analysis

One of basic issues is to investigate how are the typologies of relationships as between debt-asset acquisition and between debt-asset depletion in rural and urban cultivator households, and further to investigate whether the typologies vary across farm size groups. The present section is an attempt to investigate this issue with the National Sample Survey Organisation (NSSO) data for the years 1991-92 and 2002-03. The data availability constraints limit the analysis to only four typologies with data for the year 2002-03, where as the analysis deals with eight typologies with the data for the year 1991-92. The results given in Tables 5.25-5.27 show:

- Four Typologies of Debt-Asset Depletion: The major inferences from the analysis of 2002-03 data (Table 5.25) on this theme are:
  - In a four-typology structure, out of total sample farm households, over 80 per cent of small holders are in the typology of low debt-

Table 5.24

Compositions of Depletion of Assets and Temporal Shifts in Three Regions

Percentage Share

	LFR								*MPR*								*HPR*							
	S		*M*		*L*		*Overall*		*S*		*M*		*L*		*Overall*		*S*		*M*		*L*		*Overall*	
	II	*I*	*II*	*I*	*II*	*I*	*II*	*I*	*II*	*I*	*II*	*I*	*II*	*I*	*II*	*I*	*II*	*I*	*II*	*I*	*II*	*I*	*II*	*I*
FBA	68 (-)	85	64 (-)	80	64 (+)	49	66 (-)	76	61 (+)	57	65 (-)	83	72 (-)	99	62 (-)	65	55 (-)	69	73 (-)	76	52 (+)	17	61 (-)	68
NFBA	3 (+)	2	2 (+)	1	0 (-)	27	2 (-)	6	3 (-)	30	1 (-)	6	0	0	2 (-)	23	12 (+)	2	0 (-)	6	0	0	8 (+)	3
HHA	29 (+)	12	34 (+)	19	35 (+)	20	32 (+)	17	36 (+)	12	34 (+)	8	28 (+)	1	36 (+)	11	33 (+)	17	27 (+)	13	48 (-)	83	31 (+)	21
FA	0	0	0	0	0 (-)	3	0 (-)	1	0 (-)	1	0 (-)	3	0	0	0 (-)	2	1 (-)	12	0 (-)	5	0	0	0 (-)	8
Total	100	100	100	100	100	100	100	100	100	100	100	100	100	100	100	100	100	100	100	100	100	100	100	100

Source: Estimated by authors.

low asset depletion and low debt-high asset depletion, and the percentage share of medium and large holdings is quite low under these typologies. It is also true under high debt and low asset depletion and high debt and high depletion. The all-India sample as well as rural and urban sample presents the same typology structure.

Table 5.25

Percentage Distribution of Cultivator Households under Four Typologies: 2002-03

	Low Debt and Low Depletion	*Low Debt and High Depletion*	*High Debt and Low Depletion*	*High Debt and High Depletion*
		Rural		
S	88.44	80.30	77.24	68.37
M	11.17	18.53	21.24	28.74
L	0.39	1.17	1.52	2.89
		Urban		
S	82.95	82.99	74.62	70.53
M	15.48	13.92	22.72	27.37
L	1.57	3.09	2.66	2.10

Note: S = Small Farms;
M = Medium Farms;
L = Large Farms.

Source: Estimated by authors.

- Following the result preceded, it is also possible to infer that S farm households claim a greater share under all four typologies, and L farm households smaller share. The M farm households fall in between these two contrasting patterns smaller the farm size and higher the share in asset depletions in relation to debt 'status'.

- Eight Typologies of Debt-Asset Depletion and Acquisition: Table 5.26 presents the analysis of eight typologies of debt and asset depletion-acquisition with NSSO survey data for the year 1991-92.
  - At all-India level, the share of small farm holders is the highest in all the eight typologies, followed by medium and large holders. The difference between small and large holders is

Table 5.26

Percentage Distribution of Cultivator Households under Eight Typologies: 1991-92

	Low Debt, Low Depletion and Low Acquisition	*Low Debt, Low Depletion and High Acquisition*	*Low Debt, High Depletion and Low Acquisition*	*Low Debt, High Depletion and High Acquisition*	*High Debt, Low Depletion and Low Acquisition*	*High Debt, Low Depletion and High Acquisition*	*High Debt, High Depletion and Low Acquisition*	*High Debt, High Depletion and High Acquisition*
				India				
S	76.08	63.47	68.35	59.77	66.85	53.47	57.38	48.25
M	22.50	33.25	29.00	35.86	30.63	40.42	37.70	45.14
L	1.42	3.28	2.65	4.36	2.52	6.11	4.92	6.61
				Rural				
S	75.29	60.73	68.02	58.54	64.94	46.11	55.05	43.66
M	23.27	35.69	29.24	37.01	32.52	47.18	39.91	49.30
L	1.44	3.58	2.74	4.45	2.53	6.70	5.05	7.04
				Urban				
S	84.91	81.99	72.31	71.93	77.98	80.39	76.92	70.45
M	13.94	16.77	26.15	24.56	19.60	15.69	19.23	25.00
L	1.16	1.24	1.54	3.51	2.42	3.92	3.85	4.55

Source: Estimated by authors.

quite contrasting. For example, the percentage share of small holders under the typology of low debt-low asset depletion and acquisition is about 76 per cent, whereas the share of large holder is less than 2 per cent. This contrast is quite evident in case of all eight typologies. Further the share of small holders is lower under high debt-low depletion and high acquisition; high debt, high depletion and low acquisition; and high debt-high depletion and high acquisition typologies than under the first five typologies. But the small holders of urban areas present a different picture.

- The major inferences on the share of three rural and urban farm groups are not anything different. The share of small holder is the highest, followed by medium (with some exceptions) and large holders. For example, the share of rural small holder is about 75 per cent under low debt-low asset depletion and acquisition typology, whereas it is less than 2 per cent in case of large holders. It is equally true in case of contrast between urban small and large holders.

- An examination of results (Table 5.27) on percentage distribution of each category of cultivator households across eight typologies provides insights into the 'status' of each farm group in eight typology structure.
  - At all-India level, about 74 per cent of total S holders are in the typology of low debt-low asset depletion and acquisition typology: The remaining 26 per cent of these households are distributed across other seven typologies. In case of medium and large holdings also, the percentage share under low debt-low depletion and low acquisition typology is quite high compared to share under other seven typologies, but it is lower than the share of small holders.
  - Across three productive regions also, out of total S farm households, their share is the highest under low debt-low depletion and low acquisition category. This patter of typology varies inversely with farm size. For example, in LPR about 73 per cent of small holders are under this typology, whereas in

Table 5.27

Percentage Distribution of Cultivator Households in India and in Three Agriculture Productivity Regions under Eight Typologies: 1991-92

	Low Debt, Low Depletion and Low Acquisition	*Low Debt, Low Depletion and High Acquisition*	*Low Debt, High Depletion and Low Acquisition*	*Low Debt, High Depletion and High Acquisition*	*High Debt, Low Depletion and Low Acquisition*	*High Debt, Low Depletion and High Acquisition*	*High Debt, High Depletion and Low Acquisition*	*High Debt, High Depletion and High Acquisition*
				INDIA				
S	74.12	4.56	3.26	2.12	12.96	1.46	0.80	0.71
M	62.28	6.78	3.93	3.62	16.87	3.13	1.50	1.89
L	50.84	8.65	4.64	5.70	17.93	6.12	2.53	3.59
				LPR				
S	72.90	5.10	4.98	2.46	11.42	1.79	0.62	0.73
M	65.20	7.11	5.99	3.36	12.83	3.09	1.71	0.72
L	51.85	10.65	7.41	6.02	14.81	5.09	1.85	2.31
				MPR				
S	80.29	3.62	2.23	1.57	10.76	0.54	0.61	0.39
M	68.52	5.83	3.69	4.12	13.29	1.97	0.60	1.97
L	65.22	13.04	0.00	8.70	4.35	0.00	4.35	4.35
				HPR				
S	62.59	7.61	3.06	3.41	16.50	3.27	1.64	1.92
M	49.64	7.19	1.98	3.60	27.16	5.76	1.44	3.24
L	43.24	5.41	0.00	2.70	32.43	5.41	0.00	10.81

Note: S, M and L: As under Table Region
LPR = Low Productivity Region
MPR = Medium Productivity Region
HPR = High Productivity Region.

Source: Estimated by authors.

case of L farms it is about 52 per cent. The same contrast is observed as between small and large holders in MPR and HPR. Further, a greater percentage of small farmers of LPR are in low debt-low depletion and acquisition typology than those of HPR. Obviously the M farm households fall in between S and L farm households in their debt-asset depletion and acquisition 'behaviour.' Further, about 74 per cent of small holders are 'trapped' in the typology of low debt-low asset depletion and asset acquisition typology at all-India level.

Capital Expenditure: Debt Relations

The issue that is examined in this section is whether outstanding term loan (medium and long-term loan) has any impact on capital expenditure incurred by cultivator households of the present study. In this context, data sets used need to be made explicit. The outstanding term loan refers to a cumulative total of term loan during the period of survey, and capital expenditure to expenditure incurred during the study period on capital assets such as LD, IC, AC and FM. In this study, only the ratio of outstanding term loan per unit of capital expenditure and the rate of capital expenditure per unit of outstanding term loan are considered. These ratios are reported under A and B of Tables 5.28-5.33. Under A, the amount of outstanding term loan per unit of capital expenditure is considered, cutting across capital expenditure on different kinds of capital assets. Under B, the amount of capital expenditure per unit of term loan is considered. Logically, three possible values of A ratios could be thought of. First, the ratio is equal to one, i.e., for every unit of capital expenditure the term loan is one. Second, the rate is greater than one i.e., for every unit of capital expenditure, the average cultivator household has an outstanding loan of more than one, implying that this average household has incurred capital expenditure, even though its outstanding term loan is quite high. Third, if the ratio is less than one, this average households 'outstanding term loan status' is 'comfortable enough' to make it to spend more on capital expenditure. But quite opposite inferences could be derived from rations under B. These ratios are likely to vary across farm groups of rural and urban, and of the three regions of this study. The ratios are shown in Tables 5.28-5.33.

Table 5.28

Outstanding Term Loan Per Unit of Farm Capital Expenditure and Farm Capital Expenditure Per Unit of Outstanding Term Loan: All India

Farm Category	Ratio			
	II		I	
	A	B	A	B
Small	1.5	0.67	0.9	1.11
Medium	1.8	0.56	0.9	1.11
Large	2.5	0.40	0.5	2.00

Note: Ratio: A: Outstanding Term Loan/Capital Expenditure;
Ration B: Capital Expenditure/Outstanding Term Loan.

Source: Estimated by authors.

Table 5.29

Outstanding Term Loan Per Unit of Farm Capital Expenditure and Farm Capital Expenditure Per Unit of Outstanding Term Loan: Rural and Urban

Farm Category	Ratio							
	Rural				Urban			
	II		I		II		I	
	A	B	A	B	A	B	A	B
Small	1.5	0.67	0.9	1.11	1.0	1.00	0.7	1.43
Medium	1.9	0.53	0.9	1.11	1.5	0.67	0.7	1.43
Large	2.8	0.36	0.5	2.00	1.5	0.67	0.5	2.00

Source: Estimated by authors.

Ratio Analysis of Farm Capital Expenditure

Tables 5.28-5.30 present results on farm capital expenditure-debt relations. First, the results are analysed with ratios reported under A.

- The ratios of all groups of farms are higher during the second period than during the first period of study. This implies that an average farm household at all three levels of study—all India, rural and urban and three regions—has a greater amount of 'outstanding term loan' for every one unit of capital expenditure incurred on farm assets.

- L farms at all-India level, and rural and urban have higher amount of outstanding term loan per unit of capital expenditure than the other two farm groups. In other words, the ratios vary inversely with an increase in farm size.

  In relation to capital expenditure on farm capital assets, the L farms 'own' the highest amount of 'outstanding term loan,' and the S farms least amount of 'outstanding term loan.' But in case of all the three groups, the ratios are greater than one, excepting the ratio in case of S farms of urban area during the second period.

- Region-wise, barring the differences in farm size, the ratios are more than one in all the three regions. The ratios vary inversely with agricultural productivity status of the region, i.e., an average farm household of LPR region has the highest amount of 'outstanding term loan,' the lowest being in HPR, and in MPR the ratio falling in between.

Table 5.30

Outstanding Term Loan Per Unit of Farm Capital Expenditure and Farm Capital Expenditure Per Unit of Outstanding Term Loan in Three Regions

Region	*Ratio*			
	II		*I*	
	A	B	A	B
LPR	2.0	0.50	0.6	1.67
MPR	1.7	0.58	1.0	1.00
HPR	1.1	0.91	1.1	0.91

Source: Estimated by authors.

Ratio Analysis of Non-farm Capital Expenditure

It could be seen from Tables 5.31-5.33 that:

- At all-India level as well as at rural area, the average S farm household owes more 'outstanding term loan' per unit of non-farm capital expenditure during the second period. An average urban household small farm also owes more 'outstanding term loan' per unit of non-farm capital expenditure. At all-India level, both the M and L farms have experienced a ratio of lower outstanding term loan during the second period, whereas ratio has increased in S farm households.

Table 5.31

Outstanding Term Loan Per Unit of Non-farm Capital Expenditure and Non-farm Capital Expenditure Per Unit of Outstanding Term Loan: All India

Farm Category	*Ratio*			
	II		*I*	
	A	*B*	*A*	*B*
Small	2.4	0.43	1.1	0.91
Medium	2.3	0.43	3.1	0.32
Large	0.3	3.33	2.0	0.50

Source: Estimated by authors.

Table 5.32

Outstanding Term Loan Per Unit of Non-farm Capital Expenditure and Non-farm Capital Expenditure Per Unit of Outstanding Term Loan: Rural and Urban

Farm Category	*Ratio*							
	Rural				*Urban*			
	II		*I*		*II*		*I*	
	A	*B*	*A*	*B*	*A*	*B*	*A*	*B*
Small	2.2	0.45	0.9	1.11	3.4	0.29	1.3	0.77
Medium	2.2	0.45	1.5	0.67	2.6	0.38	6.3	0.16
Large	0.4	2.5	0.5	2.00	-	-	5.6	0.18

Source: Estimated by authors.

The S and M farm households of rural areas are found to have experienced an increase in the ratio, whereas the L farms have experienced a marginal decrease.

Between S and M farms of urban area, there is a considerable increase in the ratio of S farms during the second period, and a considerable decrease in the ratio of M farms.

Among the three regions (Table 5.33), there is a considerable increase during the second period in the ratio of outstanding term loan to capital expenditure on NFBA of S farms of both LPR and MPR. In case of M farms, the ratios have decreased in all the three regions, implying that at the time of incurring capital expenditure on NFBA, they have less of outstanding term loan in the second period than in the first period.

Table 5.33

Outstanding Term Loan Per Unit of Non-farm Capital Expenditure and Non-farm Capital Expenditure Per Unit of Outstanding Term Loan in Three Regions

Farm Category	LPR				MPR				HPR			
	II		I		II		I		II		I	
	A	B	A	B	A	B	A	B	A	B	A	B
Small	4.6	0.22	0.7	1.43	3.8	0.26	1.4	0.71	1.8	0.56	2.5	0.40
Medium	1.3	0.77	5.2	0.19	4.1	0.24	10.8	0.09	2.2	0.45	3.4	0.29
Large	0.4	2.5	0.3	3.33	-	-	-		-	-	-	-

Source: Estimated by authors.

Yet, another way of examining the results is to consider ratios reported under B. With the comparison of ratios reported under A and B, the major inference derived is: higher the ratio of term loan/capital expenditure, lower is the ratio of capital expenditure/term loan. This inference is simply the outcome of difference in ratio definitions of A and B. To illustrate this, in Table 5.28, the ratio of outstanding term loan per unit of capital expenditure is 2.5 (under ratio A) during the second period in L farms, but under B capital expenditure per unit of outstanding term loan is 0.40 for the same L farms, whereas during the first period, the corresponding ratios are 0.5 and 2.0. The implication of this result under B is that L farms in an average have incurred less capital expenditure per unit of term loan during the second period than during the first period. If ratio under A is considered for same L farm sample, loan per unit of capital expenditure is higher during the second period. But both the results would mean the same thing.

Determinants of Total Capital Expenditure

The empirical question that is addressed in this section is: what are the determinants of total capital expenditure incurred by farm households? It is logically postulated that total capital expenditure depends on total assets, total debt, household size, age of the head of the household, farm size and literacy. Further, it is also postulated that there are regional effects on total capital expenditure. The empirical results derived from all-India rural and urban samples, and regional rural and urban samples are shown in Tables 5.34-5.37.

Table 5.34

Ordering Least Squares (OLS) Regression of Total Capital Expenditure: All-India—Rural

	Elasticities/Coefficients		*Standard Error*		*t-value*		*P-value*	
Variables	*II*	*I*	*II*	*I*	*II*	*I*	*II*	*I*
ASSET	0.359	0.257	0.010	0.014	36.76	17.98	***	***
DEBT	0.035	0.069	0.004	0.006	8.34	10.97	***	***
HHS	0.235	0.199	0.015	0.022	15.26	8.85	***	***
AoH	0.092	0.109	0.026	0.036	3.56	2.99	***	***
DEDU	0.099	0.192	0.016	0.026	6.33	7.27	***	***
FS	0.435	0.469	0.034	0.050	12.61	9.28	***	***
Constant	-2.319	-1.498	0.111	0.153	-20.95	-9.82	***	***
Adjusted R Square	0.0869	0.0718						
Number of Sample	45054	21700						

Note: ASSET = Log Function of Total Assets;
DEBT = Log Function of Total Debt;
HHS = Log Function of Household Size;
AoH = Log Function of Age of Head;
DEDU = Dummy variable of literacy/illiteracy of head where literacy = 1 and illiteracy = 0;
FS = Log Function of Farm Size.

Source: Estimated by authors.

It is seen from Table 5.34 (all-India rural) that: Both during the second and first periods, total asset, total debt, household and farm size, age of the head and education status are found to have had positive and significant impact on total capital expenditure. With an increase in asset accumulation, in total debt, in size of the household, in age of the head of the family, in literacy status of the head, and in farm size, the elasticity of total capital expenditure also increases. The positive sign of coefficients of two variables *viz.*, debt and family size needs some explanation. It could be postulated that any increase in debt will have negative impact on capital expenditure. The positive impact of debt on capital expenditure appears to support the inference that availment of loan is positively supportive to increase capital expenditure. Any increase in family size could also be postulated to have negative impact on capital expenditure. But positive impact of family size on capital expenditure may imply the 'ability' of the family size to incur more capital expenditure.

In respect of results (Table 5.35) derived from total urban sample: (a) only two variables *viz.*, asset and household size have positive and significant impact on capital expenditure during the second period. (b) During the first period, asset, debt and farm size have positive and significant impact on capital expenditure.

Table 5.35

OLS Regression of Total Capital Expenditure: All-India—Urban

	Elasticities/Coefficients		*Standard Error*		*t-value*		*P-value*	
Variables	*II*	*I*	*II*	*I*	*II*	*I*	*II*	*I*
ASSET	0.451	0.442	0.034	0.041	13.28	10.89	***	***
DEBT	0.022	0.062	0.015	0.020	1.5	3.13		***
HHS	0.204	0.082	0.071	0.082	2.87	1	***	
AoH	-0.153	-0.188	0.123	0.154	-1.25	-1.22		
DEDU	0.067	0.068	0.078	0.100	0.86	0.68		
FS	0.200	0.471	0.146	0.192	1.37	2.45		**
Constant	-1.933	-1.418	0.470	0.569	-4.11	-2.49	***	**
Adjusted R Square	0.0798	0.0872						
Number of Sample	3241	2335						

Source: Estimated by authors.

In case of results (Table 5.36) on all three region rural samples: (a) explanatory variables such as asset, debt, household size, education and farm size have positively and significantly impacted capital expenditure during both the periods. (b) The coefficients of dummy variable for MPR are both negative and significant during both the periods. In case of HPR, the coefficient is both positive and significant only during the second period.

With respect to urban samples of the three regions (Table 5.37): (a) the coefficients of only two variables *viz.*, asset and household size are significant and positive. (b) During the first period, debt has also exerted positive and significant impact on capital expenditure of urban farm households. Only in case of MPR, the regional effect is found to be both negative and significant during the second period.

Table 5.36

OLS Regression of Total Capital Expenditure: Three Agricultural Productivity Regions: Rural

	Elasticities/Coefficients		*Standard Error*		*t-value*		*P-value*	
Variables	*II*	*I*	*II*	*I*	*II*	*I*	*II*	*I*
ASSET	0.361	0.279	0.013	0.018	27.09	15.18	***	***
DEBT	0.022	0.065	0.005	0.009	4.39	6.9	***	***
HHS	0.287	0.212	0.021	0.032	13.51	6.71	***	***
AoH	0.067	0.042	0.036	0.057	1.89	0.73	*	
DEDU	0.078	0.141	0.020	0.035	3.8	4.01	***	***
FS	0.443	0.477	0.058	0.070	7.6	6.77	***	***
DHPR	0.117	0.031	0.041	0.063	2.84	0.49	***	
DMPR	-0.328	-0.404	0.025	0.047	-13.26	-8.66	***	***
Constant	-2.082	-1.294	0.139	0.235	-14.96	-5.51	***	***
Adjusted R Square	0.1048	0.098						
Number of Sample	24707	11126						

Source: Estimated by authors.

Table 5.37

OLS Regression of Total Capital Expenditure: Three Agricultural Productivity Regions: Urban

	Elasticities/Coefficients		*Standard Error*		*t-value*		*P-value*	
Variables	*II*	*I*	*II*	*I*	*II*	*I*	*II*	*I*
ASSET	0.489	0.535	0.050	0.063	9.72	8.5	***	***
DEBT	0.008	0.080	0.019	0.029	0.41	2.78		***
HHS	0.259	0.278	0.101	0.111	2.56	2.51	**	**
AoH	-0.255	-0.246	0.157	0.193	-1.62	-1.27		
DEDU	-0.015	-0.116	0.092	0.138	-0.16	-0.84		
FS	0.285	0.065	0.222	0.291	1.28	0.22		
DHPR	0.152	0.047	0.145	0.171	1.05	0.27		
DMPR	-0.383	-0.139	0.118	0.154	-3.24	-0.9	***	
Constant	-1.601	-2.203	0.677	0.755	-2.37	-2.92	**	***
Adjusted R Square	0.1112	0.1226						
Number of Sample	1735	1002						

Source: Estimated by authors.

6 NCAER Data Based Analysis of Asset

Liability Portfolio, Financial Inclusion and Capital Expenditure

The preceding chapters 3-5 address the issues of asset-liability portfolio of farmer households, the determinants of financial inclusion/exclusion of these households, and structure and determinants of capital expenditure, using National Sample Survey (NSS) household level data. The present chapter addresses four major issues, drawing household level data set from the data pool of the National Council of Applied Economic Research (NCAER) for the years 1998-99 and 2005-06:

- First, how is the asset-liability portfolio structure of farmer households across India?
- Second, what are the determinants of non-farm business assets?
- Third, what factors would determine financial inclusion and exclusion of farmer households from getting access to institutional sources of credit?
- Fourth, what are the determinants of capital expenditure incurred by the sample farmer households?

NCAER has collected data through four rounds of Additional Rural Incomes Survey (ARIS) and Rural Economic and Demographic Survey (REDS) between 1971 and 2006. These four rounds of data collected by NCAER would constitute almost panel data of over three decades, with some additions/replacements to and omissions from this panel data from villages spread across the country. Details on methodology of this longitudinal data collection are in NCAER (1975; 1986; 1987), Deninger *et al.* (2009), Jain (2011) and Vashishtha (n.d.).

A brief reference to dataset used in this study for 1998-99 and 2005-06 is in order. First, even though over 7474 sample households in 1998-99 and 9500 households during 2005-06 were surveyed by NCAER, data set for the present study has been extracted, depending on the availability of complete data set for each household. It is expected that the estimates are not likely to be totally unbiased. However, the major inferences drawn from the analysis are not likely to be affected due to this choice of data set. Second, there have been some additions to the number of aspects/variables on which data have been collected by NCAER during different rounds. For example in 2005-06 round, data on sample household participation in governance are also included. Third, in general, NCAER unit (household) level data would provide micro characteristics of sample households and their interactions with environments village/district/state/national (particularly with respect to national policies). The household schedules would provide data on assets, liabilities, incomes by source, agricultural inputs and output. It would also provide household members demographic characteristics including educational attainment, school enrolment and work participation. The empirical results are organised in two parts: Part A: 1998-99 and Part B: 2005-06.

PART A: ANALYSIS OF RESULTS: 1998-99

Composition of Asset Portfolio

Empirical evidences on asset-portfolio structure are shown in Tables 6.1-6.8.a.

It could be observed from Table 6.1 that farm business assets (FBA) would form the highest share of over 80 per cent of total assets of farmer households, followed by household assets (HHA, 13.1%), livestock (2.3%), and NFBA forming far less than 1 per cent implying lack of asset-employment-income diversification in rural areas. Not much of a difference is observed in the asset portfolio of three farm groups, excepting the share of livestock asset in total asset of small farmer households exceeds marginally that of large farmer households.

Table 6.1

Composition of Assets: All India: 1998-99

Percentage Share

Assets	*S*	*M*	*L*	*Overall*
FBA	85.8	82.4	86.6	84.4
NFBA	0.2	0.4	0.1	0.2
HHA	11.8	14.7	11.4	13.1
LIVESTOCK	2.2	2.6	1.9	2.3
Total Assets	100	100	100	100

Note: FBA = Farm Business Assets;
NFBA = Non-farm Business Assets;
HHA = Household Assets;
LIVESTOCK = Livestock consists of agricultural and non-agricultural purpose;
S = Marginal and Small Farms (<2 ha);
M = Semi-medium and Medium Farms (>= 2 and <10 ha);
L = Large Farms (>= 10 ha)
Sample observations: S = 2516, M = 1979, L = 177, Total obs. = 4672

Source: Estimated by authors.

The results on the disaggregative analysis of farm business assets are given in Table 6.2. In overall, land alone constitutes 94 per cent of total assets, and the remaining 6 per cent shared equally by irrigation capital and farm machinery. The overwhelmingly dominant share of land is observed in case of all the three farm groups, excepting farm machinery with large

Table 6.2

Disaggregative Analysis of Portfolio of Farm Business Assets: All India: 1998-99

Percentage Share

Components	*S*	*M*	*L*	*Overall*
La	97	91	89	94
IC	2	4	3	3
FM	1	4	8	3
OFA	0	1	0	0
Total	100	100	100	100

Note: La = Land of Farm Business;
IC = Irrigation Capital;
FM = Farm Machinery for Farm Business;
OFA = Other Farm Assets.

Source: Estimated by authors.

farmer households claim share of 8 per cent which is far higher than the combined share of both S and M farms in farm machinery.

The disaggregative results on composition of HHA are shown in Table 6.3. All the three farm groups exhibit the dominance of share of buildings, followed by household durable assets. As between small and large holding families, household durable (HHD) assets form a greater share in large farmer households whereas buildings constitute a greater share in small farmer households.

Table 6.3

Disaggregative Analysis of Portfolio of Household Assets: All India: 1998-99

Percentage Share

Components	*S*	*M*	*L*	*Overall*
B	81	75	70	76
HHPROP	5	6	6	6
HHD	14	19	24	18
Total	100	100	100	100

Note: B = Buildings;
HHPROP = Property for Household;
HHD = Household Durable Goods.

Source: Estimated by authors.

With respect to composition of financial assets (Table 6.4), gold turns out to command the highest share in all the farm groups, followed by deposits and savings. But the investment in shares is quite insignificant, and asset holdings in the form of currency would form about 11 per cent.

The aggregative asset component analysis of farmer households of three agricultural productivity regions (Table 6.5) suggests the dominance (over 85%) of FBA in all the farm groups of the three regions, followed by HHA and livestock. However the percentage share of livestock in total assets is higher in low-productivity region (LPR) than in other two regions. This is perhaps due to dependence of LPR farmer household on livestock for their livelihood. Added to all these is the insignificance of NFBA in total asset holding in all the three regions, implying lack of development of non-farm economy in rural areas.

Table 6.4

Disaggregative Analysis of Portfolio of Financial Assets: All India: 1998-99

Percentage Share

Financial Assets	*S*	*M*	*L*	*Overall*
Deposits	19	16	22	18
Savings	17	18	21	18
Shares	0	1	2	1
Gold	53	54	45	53
Currency	10	11	11	11
Total	100	100	100	100

Source: Estimated by authors.

Table 6.5

Composition of Assets: Three Agricultural Productivity Regions: 1998-99

Percentage Share

Assets	*LPR*				*MPR*				*HPR*			
	S	*M*	*L*	*Overall*	*S*	*M*	*L*	*Overall*	*S*	*M*	*L*	*Overall*
FBA	91	84	85	86	92	78	77	89	70	85	91	85
NFBA	0	0	0	0	0	0	0	0	0	1	0	0
HHA	7	13	11	11	7	20	22	10	26	13	8	13
LIVESTOCK	1	3	4	3	1	2	1	1	3	2	1	2
Total Assets	100	100	100	100	100	100	100	100	100	100	100	100

Note: LPR = Low-productivity regions (Madhya Pradesh (MP), Rajasthan);
MPR = Medium-productivity regions (Uttar Pradesh (UP); Bihar);
HPR = High-productivity regions (Punjab, Tamil Nadu);
Other definitions as under Table 6.1;
Sample observations: S = 1157, M = 1054, L = 109, Total obs. = 2320.

Source: Estimated by authors.

Component analysis (Table 6.6) of FBA in total FBA suggests the dominant share of land (much more so in case of S farms), followed by the share of irrigation capital in LPR and of farm machinery in HPR.

With respect to disaggregative analysis (Table 6.7) of HHA portfolio, building component obviously forms the highest share, followed by HHD assets. As between large and small farmer households, the dominance of building share in small farmer households is much more than in large farmer households.

Table 6.6

Disaggregative Analysis of Portfolio of Farm Business Assets: Three Agricultural Productivity Regions: 1998-99

Percentage Share

Components	*LPR*				*MPR*				*HPR*			
	S	*M*	*L*	*Overall*	*S*	*M*	*L*	*Overall*	*S*	*M*	*L*	*Overall*
La	96.8	91.2	91.5	93.2	99.6	95.4	92.2	98.6	84.9	91.3	88.5	89.8
IC	2.9	6.1	3.5	4.6	0.2	1.1	0.4	0.3	11.6	3.1	2.2	3.6
FM	0.2	2.3	4.7	2.0	0.3	3.3	7.2	1.0	3.2	5.4	9.2	6.4
OFA	0.1	0.4	0.2	0.2	0.0	0.2	0.2	0.1	0.4	0.2	0.1	0.2
Total	100	100	100	100	100	100	100	100	100	100	100	100

Note: Definitions of abbreviations as under Tables 6.1 and 6.2.
Source: Estimated by authors.

Table 6.7

Disaggregative Analysis of Portfolio of Household Assets: Three Agricultural Productivity Regions: 1998-99

Percentage Share

Components	*LPR*				*MPR*				*HPR*			
	S	*M*	*L*	*Overall*	*S*	*M*	*L*	*Overall*	*S*	*M*	*L*	*Overall*
B	86.7	81.9	74.8	81.7	92.1	87.5	86.6	89.9	80.9	70.3	72.1	72.9
HHPROP	1.1	1.8	6.3	2.4	0.2	0.1	0.0	0.1	4.2	5.0	6.1	5.0
HHD	12.2	16.3	18.8	15.8	7.7	12.4	13.4	10.0	14.9	24.7	21.8	22.0
Total	100	100	100	100	100	100	100	100	100	100	100	100

Note: Definitions of abbreviations as under Tables 6.1 and 6.2.
Source: Estimated by authors.

The results on component analysis of financial assets portfolio of the three regions are given in Table 6.8.a. In all the three regions, overall gold claims the highest share, but it is less in HPR region than in other two regions. With respect to deposits and savings, the highest share is found to be in HPR, followed by MPR and LPR. Financial assets in the form of insurance also constitute about 10 per cent in LPR and MPR, but 18 per cent in HPR. In overall, it is surprising to observe that farmers of LPR hold a greater proportion of their financial assets in the form of currency than those of MPR and HPR. Excepting some insignificant percentage share held in the form of shares in HPR, the participation of households of LPR and MPR in share market is almost zero.

Table 6.8.a

Disaggregative Analysis of Portfolio of Financial Assets: Three Agricultural Productivity Regions: 1998-99

Percentage Share

Financial Assets	*LPR*				*MPR*				*HPR*			
	S	*M*	*L*	*Overall*	*S*	*M*	*L*	*Overall*	*S*	*M*	*L*	*Overall*
Deposits	9	11	13	11	9	10	19	10	12	19	28	18
Savings	11	14	10	13	20	20	18	20	16	23	34	23
Shares	0	0	0	0	0	0	0	0	0	1	1	1
Gold	60	58	65	59	61	65	60	63	55	42	25	43
Currency	20	17	13	17	9	5	3	7	17	15	12	15
Insurance	9	11	13	11	9	10	19	10	12	19	28	18
Total	100	100	100	100	100	100	100	100	100	100	100	100

Source: Estimated by authors.

Related to the analysis of asset portfolio structure is to examine the relationship between NFBA and explanatory variables such as age of the household head, literacy level of head, net worth of the household, farm size and household size. It could be observed from Table 6.8.b that: (a) education

Table 6.8.b

Bootstrap Tobit Regression of Non-Farm Business Asset: All India: 1998-99

Variables	*Elasticities/Coefficients*	*Standard Error*	*t-value*	*P-value*
AoH	1.89	2.94	0.64	
DEDU	8.57	2.27	3.77	***
NW	1.08	0.82	1.31	
FS	-9.02	2.45	-3.67	***
HHS	5.31	2.14	2.48	**
Constant	-52.48	14.82	-3.54	***
Log Likelihood	-537.68			
Number of Sample	895			

Note: AoH = Log Function of Age of Head;
DEDU = Dummy variable of literacy/illiteracy of head where literacy = 1 and illiteracy = 0;
NW = Log Function of Net Worth (Net Worth = Total assets – Total outstanding debt);
FS = Log Function of Farm Size;
HHS = Log Function of Household Size;
'***' = Level of significance 1%; '**' = Level of significance 5%;
'*' = Level of significance 10%;
Since almost more than 90 per cent observations of dependent variable are 0 and the rests are continuous, so a censored model i.e., Tobit has been applied in the place of OLS with lower limit as 0.
Bootstraps have been performed for each regression and they have been replicated 1000 times.

Source: Estimated by authors.

of the head of the household and household size are found to have positive impact on NFBA acquisition. (b) Farm size is found to have negative impact on NFBA accumulation. This perhaps could be explained in terms of farmer households engagement more in FB activities with less of attention to NFB activities. (c) Both age of the household head and household net worth are found to have positive impact on NFBA accumulation, but the coefficients are not statistically significant.

Compositional Analysis of Liability Portfolio

The results on compositional analysis of liability portfolio are given in Tables 6.9-6.16.

Overall, 74 per cent of outstanding loans of sample farmer households has been from institutional sources, and 26 per cent from non-institutional sources. The percentage share of loans from institutional sources in case of large farmer households has been greater by 10 percentage points than in the case of small farmers. In other words, small farmers depend more on non-institutional sources than large farmers (Table 6.9).

Table 6.9

Compositions of Outstanding Loans: All India: 1998-99

Percentage Share

Components	*S*	*M*	*L*	*Overall*
Institutional Loans	68	77	78	74
Non-institutional Loans	32	23	22	26
Total	100	100	100	100

Note: Sample observations: S = 402, M = 446, L = 47, Total obs. = 895.

Source: Estimated by authors.

With respect to purpose-wise availment of loans, FB in general claims 61 per cent, NFB 9 per cent, household assets 3 per cent and the miscellaneous 27 per cent. As between large and small farmer households, the percentage share of loan availment for farm business is quite high among large farmer households, and whereas in case of small farmer households, the percentage share for NFB activity is at least 7, it is almost nil in case of large farmer households (Table 6.10).

Table 6.10

Component Analysis of Loan Availment: All India: 1998-99

% Share of Total Outstanding Loan

Components	*S*	*M*	*L*	*Overall*
FB	51	64	74	61
NFB	7	11	0	9
HHD	3	3	3	3
Others	39	22	23	27
Total	100	100	100	100

Source: Estimated by authors.

It could be observed from Tables 6.11 and 6.12 that in overall, FB activity claims the highest share in total loan availed from institutional sources, whereas miscellaneous purpose of loan claims the highest share in the total loan availed from non-institutional sources. Across farm groups, large farmer group has availed the highest share of loan for farm business from both the sources, and in case of small farmer households, the share of miscellaneous purpose of loan from institutional and non-institutional sources is as high as 31 per cent and 57 per cent respectively. This perhaps implies quite a high percentage of loan availed by small farmers for meeting, consumption, social and other family needs.

Table 6.11

Component Analysis of Institutional Loan Availment: All India: 1998-99

% Share of Total Outstanding Loan

Components	*S*	*M*	*L*	*Overall*
FB	58	70	73	67
NFB	9	14	0	11
HHD	2	3	0	2
Others	31	13	27	19
Total	100	100	100	100

Note: FB = Farm Business;
NFB = Non-farm Business;
HHD = Household Durables.

Source: Estimated by authors.

Table 6.12

Component Analysis of Non-institutional Loan Availment: All India: 1998-99

% Share of Total Outstanding Loan

Components	*S*	*M*	*L*	*Overall*
FB	35	45	76	44
NFB	3	1	0	2
HHD	5	2	13	4
Others	57	52	11	50
Total	100	100	100	100

Source: Estimated by authors.

All three farm groups of HPR have availed a greater share of their outstanding loan from institutional sources, whereas the farm groups of LPR in overall have availed about one-third of their outstanding loan from non-institutional sources and small farmer households of LPR about two-third from non-institutional sources. This would very much suggest policy directions for financial inclusion of small farmers of LPR and MPR (Table 6.13).

Table 6.13

Compositions of Outstanding Loans in Three Regions: 1998-99

Percentage Share

	LPR				*MPR*				*HPR*			
	S	*M*	*L*	*Overall*	*S*	*M*	*L*	*Overall*	*S*	*M*	*L*	*Overall*
IL	38	69	72	65	0	0	N/A	0	79	93	100	87
NIL	62	31	28	35	100	100	N/A	100	21	7	0	13
Total	100	100	100	100	100	100	N/A	100	100	100	100	100

Note: IL = institutional loans;
NIL = Non-institutional loans;
Sample observations: S = 201, M = 304, L = 33, Total obs. = 538.

Source: Estimated by authors.

In overall, out of the total loan availment, over 70 per cent is found to be for FB activities in case of both LPR and HPR. But there are farm-group-wise differences as between small and large farmer households. The small farmers of LPR have availed only 34 per cent of total loan for FB activities, whereas in the case of large farmer households, the corresponding percentage would come to 100. Small farmer households of LPR have availed 8 per cent for NFB activities, and 57 per cent for miscellaneous/other purposes (Table 6.14).

Table 6.14

Component Analysis of Loan Availment in Regions: 1998-99

Percentage Share in Total Outstanding Loan

Component	*LPR*				*HPR*		
	S	*M*	*L*	*Overall*	*M*	*L*	*Overall*
FB	34	75	87	71	71	100	80
NFB	8	2	0	2	0	0	0
HHD	1	2	4	3	5	0	4
Others	57	21	9	24	24	0	16
Total	100	100	100	100	100	100	100

Note: Due to problems of data, estimates are not made in case of MPR.
Source: Estimated by authors.

How are the results on component analysis between IL and NIL availment in regions (Tables 6.15 and 6.16). Large sample farmers of HPR have availed loans from institutional sources only for FB activities, and 77 per cent of loan availed from non-institutional sources is also for FB activities in LPR. In case of small farmer households of LPR, 51 per cent of loan availed from institutional sources is for FB activities, whereas in HPR the corresponding percentage would come to 94. From non-institutional sources, a greater percentage of loan availed (73%) by small farmer household of LPR is for miscellaneous purposes, but in HPR the

Table 6.15

Component Analysis of Institutional Loan Availment in Regions: 1998-99

Percentage Share in Total Outstanding Loan

Component	*LPR*				*HPR*			
	S	*M*	*L*	*Overall*	*S*	*M*	*L*	*Overall*
FB	51	89	91	86	94	71	100	81
NFB	17	2	0	3	0	0	0	0
HHD	0	2	1	1	3	5	0	4
Others	32	7	8	10	3	24	0	15
Total	100	100	100	100	100	100	100	100

Source: Estimated by authors.

corresponding percentages would come to 26. These results would establish that purpose of loan availment varies with regions and farm groups. This is one dimension of difference in the credit propensity of farm groups as between different agricultural productivity regions.

Table 6.16

Component Analysis of Non-institutional Loan Availment in Regions: 1998-99

Percentage Share in Total Outstanding Loan

Component	*LPR*				*HPR*			
	S	*M*	*L*	*Overall*	*S*	*M*	*L*	*Overall*
FB	23	44	77	43	74	73	0	74
NFB	2	0	0	1	0	0	0	0
HHD	2	4	13	5	0	0	0	0
Others	73	52	10	51	26	27	0	26
Total	100	100	100	100	100	100	100	100

Source: Estimated by authors.

Relationship between Credit Propensity and Share of Assets: 1998-99

Using results from polynomial regression, Figure 6.1 explores the relationship between credit availed per unit of cultivated area and the percentage share of four types of assets *viz.*, FBA, NFBA, HHA and livestock capital assets in total assets.

- It could be observed that the percentage share of FBA is constant for S farms, marginally decreasing for M farms and increasing for L farms with an increase in propensity to borrow.
- The percentage share of HHA initially increases and then decreases for S farms. For M farms, percentage share of HHA marginally increases, but it marginally decreases for L farms.
- Percentage share of NFBA and livestock remains constant over credit propensity across all farms. The exceptions are in the case of M farms for which the percentage share of NFBA increases marginally, and in case of L farms, percentage share of livestock initially increases marginally, and then decreases.

Figure 6.1

Relationship between Credit Propensity and Share of Assets across India in 1998-99

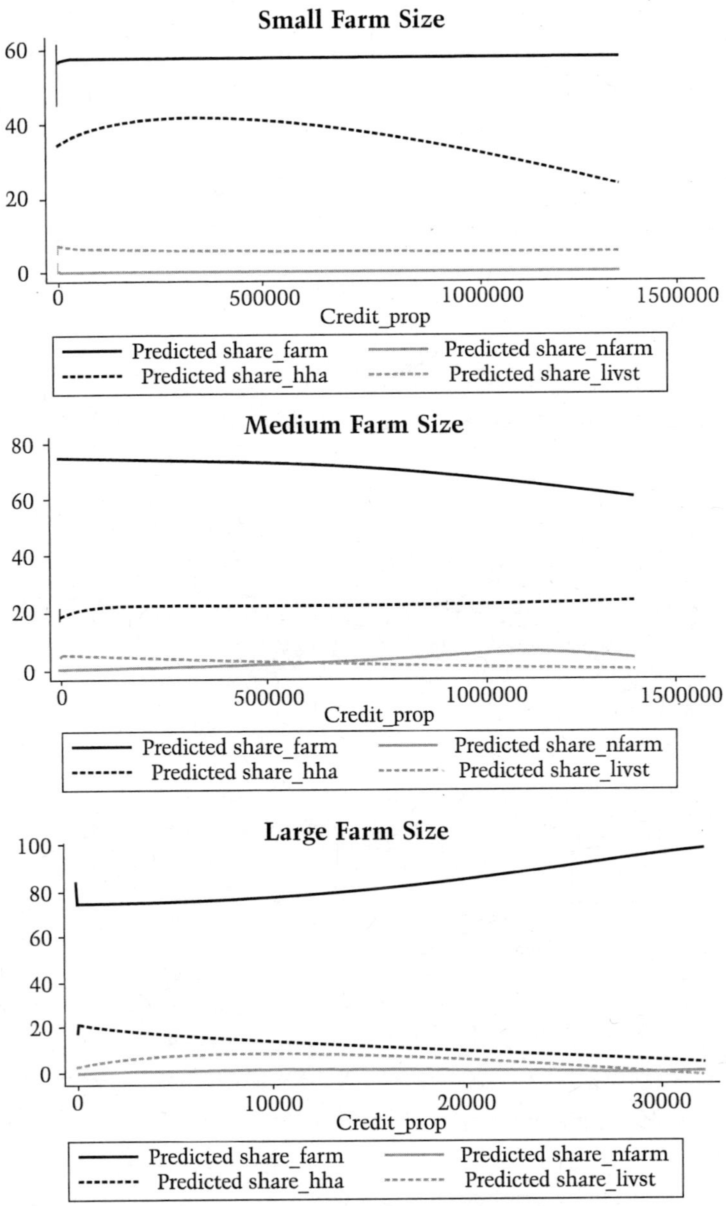

Note: X-axis = Credit Propensity = Total outstanding loan per unit land area cultivated.
Y-axis = Percentage share of four types of assets in total assets.

- The orders of percentage share of assets are almost similar for all the farm groups where FBA is the highest in the order, followed by HHA, livestock and NFBA.
- The concluding observation for M farms is that as credit propensity to borrow increases, percentage share of FBA decreases, with an increase in percentage share of HHA and NFBA. In case of large farms, percentage share of FBA increases, with an increase in credit propensity to borrow, followed by decline in percentage share of HHA and livestock.

Since FBA commands higher share in the total value of assets, an examination of relationship between FBA, IL and NIL is in order. The results are represented in Figure 6.2.

Figure 6.2

Relationship between FBA, IL and NIL in India: 1998-99

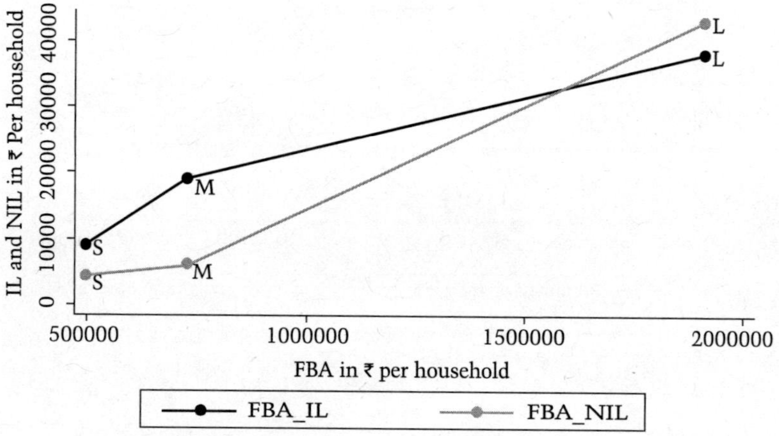

Note: X-axis = FBA in ₹ per household;
Y-axis = IL and NIL in ₹ per household.

- Black line (FBA_IL) means the relationship between FBA and IL in 1998-99 for India. Grey line (FBA_NIL) means the relationship between FBA and NIL in 1998-99 for India. S, M and L mean small, medium and large farms.
- As black line goes from S farms to M farms, FBA increases along with IL. As black line goes from M to L farms, FBA still increases along with IL but the intensity is lower than earlier.

- As grey line goes from S farms to M farms, FBA increases marginally along with NIL. As grey line goes from M to L farms, FBA increases along with NIL with intensity higher than earlier.
- When we compare both the black and grey lines together, we see that as both IL and NIL increase FBA increases, but the intensity differs.

Determinants of Financial Inclusion/Exclusion

In the previous section, the relationship between propensity to borrow and the share of four types of assets has been examined. In this section the focus is on analysing the determinants of financial inclusion/exclusion, and the results are given in Table 6.17. For the purpose of analysing determinants of financial inclusion/exclusion, bootstrap logistic regression model is used. A brief exposition of this regression model is in order.

The basic idea of bootstrapping is that inference about a population from sample data can be modelled by resampling the sample data. As the population is unknown, the true error in a sample statistic against its population value is unknowable. In bootstrap resamples, the 'population' is in fact the sample and this is known; hence the quality of inference from resample data which is becoming 'true' sample is measurable.

We do not know the population size. We only work with sample and try to infer about population. So when we perform estimation then the results of coefficient, variance and confidence interval are corresponding to that sample's result and not belong to population. Now if we need to know that the results belong to population, then we need to resample the sample size as many as possible and then only we get a huge sample set which comes from resampling which may be equivalent to true population. It has been deduced that by resampling the sample 100 times we get stable coefficients of population, by resampling 1000 times we get stable variance and by resampling 10,000 times we get stable confidence interval. In the present analysis, 1000 times bootstrapping has been done.

It could be observed from Table 6.17 that:

- Only two explanatory variables are significant. With an increase in the age of household head, the likelihood of financial inclusion increases. Similarly, an increase in farm size would increase the financial inclusion of household.

- It is appropriate to postulate that education level of the head, value of FB and NFB assets, value of livestock asset and of household assets tend to have positive impact on financial inclusion. In other words, an increase in the value of these explanatory variables would increase the likelihood of financial inclusion of these households, and a decrease in the value of these would increase the likelihood of financial exclusion. Even though the coefficients of these explanatory variables are statistically insignificant, they have the positive signs as postulated.

Table 6.17

Bootstrap Logistic Regression of Financial Inclusion: All India: 1998-99

Variables	*Odds Ratio*	*Standard Error*	*t-value*	*P-value*
EoH	1.032293	0.024566	1.34	
AoH	1.013403	0.0060875	2.22	**
FBA	1.000000	4.37e-08	0.65	
NFBA	1.000022	0.0000183	1.20	
HHA	1.000001	5.83e-07	0.86	
LIVESTOCK	1.000006	4.24e-06	1.36	
FS	1.305281	0.1861925	1.87	*
Log Likelihood	-575.75			
Number of Sample	895			

Note: EoH = Education of Head;
AoH = Age of Head;
FBA = Farm Business Assets;
NFBA = Non-Farm Business Assets;
HHA = Household Assets;
FS = Farm Size;
'***' = Level of significance 1%;
'**' = Level of significance 5%;
'*' = Level of significance 10%.

Source: Estimated by authors.

Aggregative and Disaggregative Analysis of Capital Expenditure Pattern

The results shown in Tables 6.18-6.23 provide empirical insights into capital expenditure pattern of farmer households during the survey period.

The highest share (54%) of capital expenditure incurred by large farmer households is towards farm business capital, followed by capital expenditure on HHA, like building, and livestock capital asset.

In case of small farmer households, the highest share (49%) is towards livestock capital assets followed by HHA and FBC. Only medium and small farmer households are found to have incurred some amount to acquire non-farm business capital assets (Table 6.18).

Table 6.18

Composition of Capital Expenditure: All India: 1998-99

Percentage Share

Components	*S*	*M*	*L*	*Overall*
FBC	19	25	54	29
NFBC	2	5	0	3
HHC	29	31	24	29
LivestockC	49	40	22	39
Total	100	100	100	100

Note: FBC = Capital Expenditure on Farm Business;
NFBC = Capital Expenditure on Non-Farm Business;
HHC = Capital Expenditure on Household Buildings and Land;
LivestockC = Capital Expenditure on Livestock;
Sample observations: S = 531, M = 486, L = 65, Total obs. = 1082.

Source: Estimated by authors.

With respect to composition of farm business capital assets, the highest share has gone to purchase/development of land, followed by expenditure towards irrigation capital and farm machinery in case of small farmer households. In case of large farmer households, the highest share has been towards purchase of farm machinery, followed by irrigation capital and land purchase/development. In overall, the highest share has been claimed by irrigation capital, followed by farm machinery and land (Table 6.19).

Table 6.19

Composition of Farm Business Capital Expenditure: All India: 1998-99

Percentage Share

Components	*S*	*M*	*L*	*Overall*
La	43	27	21	27
IC	33	36	35	35
FM	16	33	40	33
OFA	8	4	5	5
Total	100	100	100	100

Note: La = Farm Capital Expenditure on Land
IC = Farm Capital Expenditure on Irrigation Capital
FM = Farm Capital Expenditure on Farm Machinery
OFA = Farm Capital Expenditure on Others.

Source: Estimated by authors.

A disaggregate analysis of household capital expenditure suggests that the highest share (66%) is towards buildings, followed by expenditure (34%) on plots of land (Table 6.20).

Table 6.20

Composition of Household Capital Expenditure: All India: 1998-99

Percentage Share

Components	*S*	*M*	*L*	*Overall*
BC	66	85	86	81
HHPROPC	34	15	14	19
Total	100	100	100	100

Note: BC = Household Capital Expenditure on Buildings;
HHPROPC = Household Capital Expenditure on Plots;

Source: Estimated by authors.

Region-wise analysis (Table 6.21) of capital expenditure pattern suggests that large farmer households have incurred the highest amount of their capital expenditure on acquisition of farm business capital assets in all the three regions, whereas the small farmer households have incurred the smallest share on FBC in their total capital expenditure, and the highest share for acquisition of livestock capital in both LPR and MPR. This establishes the importance of livestock capital asset for small farmer households.

Table 6.21

Compositions of Capital Expenditure in Three Regions: 1998-99

Percentage Share

Components	*LPR*				*MPR*				*HPR*			
	S	*M*	*L*	*Overall*	*S*	*M*	*L*	*Overall*	*S*	*M*	*L*	*Overall*
FBC	29	24	54	33	19	25	49	26	10	23	63	29
NFBC	0	0	0	0	1	0	0	0	0	0	0	0
HHC	16	22	16	19	23	34	18	29	54	48	13	41
LivestockC	55	54	30	47	58	41	33	45	37	30	24	29
Total	100	100	100	100	100	100	100	100	100	100	100	100

Note: Definitions as under Table 6.18;
Sample observations: S = 202, M = 273, L = 35, Total obs. = 510.
Source: Estimated by authors.

Component (disaggregative) analysis (Table 6.22) of farm capital expenditure pattern suggests that the highest share has been claimed by irrigation capital in LPR of all three farm groups and highest share in MPR of both small and medium farmer households. Farm machinery claims the highest share (55%) of capital expenditure incurred by large farmer households of HPR region (Table 6.22).

Table 6.22

Components of Farm Capital Expenditure in Three Regions: 1998-99

Percentage Share

Components	*LPR*				*MPR*				*HPR*			
	S	*M*	*L*	*Overall*	*S*	*M*	*L*	*Overall*	*S*	*M*	*L*	*Overall*
La	29	36	34	34	24	15	80	31	99	41	11	30
IC	66	52	55	55	42	43	17	37	0	28	20	23
FM	5	8	2	5	34	39	3	30	1	32	55	41
OFA	0	4	9	6	0	3	0	2	0	0	14	6
Total	100	100	100	100	100	100	100	100	100	100	100	100

Note: Definitions as under Table 6.19.
Source: Estimated by authors.

Component analysis (Table 6.23) of household capital expenditure reveals that in overall, over 73 per cent of total expenditure is towards buildings.

Table 6.23

Components of Household Capital Expenditure in Three Regions: 1998-99

Percentage Share

Components	LPR				MPR				HPR			
	S	M	L	Overall	S	M	L	Overall	S	M	L	Overall
BC	68	73	72	73	63	87	91	81	99	92	27	89
HHPROPC	32	27	28	27	37	13	9	19	1	8	73	11
Total	100	100	100	100	100	100	100	100	100	100	100	100

Note: Definitions as under Table 6.20.
Source: Estimated by authors.

Determinants of Capital Expenditure

Using the bootstrap ordinary least squares (OLS) regression model, the empirical results on determinants of capital expenditure incurred by farmer households are given in Table 6.24.

Table 6.24

Bootstrap OLS Regression of Total Capital Expenditure: All India: 1998-99

Variables	Elasticities/Coefficients	Standard Error	t-value	P-value
ASSET	0.507	0.148	3.43	***
DEBT	0.100	0.042	2.40	**
HHS	1.308	0.242	5.41	***
AoH	-0.612	0.364	-1.68	*
DEDU	0.267	0.244	1.09	
FS	0.667	0.416	1.60	
Constant	-1.354	2.029	-0.67	
Adjusted R Square	0.1098			
Number of Sample	895			

Note: ASSET = Log Function of Total Assets;
DEBT = Log Function of Total Debt;
HHS = Log Function of Household Size;
AoH = Log Function of Age of Head;
DEDU = Dummy variable of literacy/illiteracy of head where literacy = 1 and illiteracy = 0;
FS = Log Function of Farm Size.

Source: Estimated by authors.

It is postulated that level of debt, household size, age of the head of the household, literacy and farm size tend to impact positively capital expenditure. A word of clarification is needed on positive impact of debt on capital expenditure. It is expected that in average, ignoring the farm size and level of capital expenditure tends to be positively impacted by the level of borrowing.

The results validate the postulated relationship between dependent and independent variables, excepting coefficient sign of the age of the household head. However, the coefficient is statistically insignificant.

PART B: ANALYSIS OF RESULTS: 2005-06

Composition of Asset Portfolio

Empirical results on asset portfolio of sample households during 2005-06 are shown in Tables 6.25-6.32

Overall, FBA forms the dominant share (88%) of total assets followed by HHA (7%), NFBA (4%) and livestock forming a low percentage of one. The dominant share of FBA is also observed among three farm groups, excepting the share is higher in S and M farms than in L farms. NFBA commands a higher share (17%) in case of large farms, perhaps indicative of these farms moving more towards NFB activities (Table 6.25).

Table 6.25

Composition of Assets: All India: 2005-06

Percentage Share

Assets	*S*	*M*	*L*	*Overall*
FBA	89	88	76	88
NFBA	2	5	17	4
HHA	8	6	6	7
LIVESTOCK	1	1	1	1
Total Assets	100	100	100	100

Note: FBA = Farm Business Assets;
NFBA = Non-farm Business Assets;
HHA = Household Assets;
LIVESTOCK = Livestock consists of agricultural and non-agricultural purpose;
S = Marginal and Small Farms (<2 ha);
M = Semi-medium and Medium Farms (>= 2 and <10 ha);
L = Large Farms (>= 10 ha);
Sample observations: S = 2832, M = 862, L = 48, Total obs. = 3742.

Source: Estimated by authors.

The results on disaggregative components of FBA (Table 6.26) suggest that: land capital claims the highest share of over 95 per cent, and in case of L farms, the share of both IC and FM is higher than in the case of S and M farms in whose case the share of land capital is higher.

Table 6.26

Disaggregative Analysis of Portfolio of Farm Business Assets: All India: 2005-06

Percentage Share

Components	*S*	*M*	*L*	*Overall*
La	98.6	97.4	95.4	98.0
IC	0.3	0.3	0.5	0.3
FM	0.7	1.9	3.8	1.3
OFA	0.4	0.4	0.3	0.4
Total	100	100	100	100

Note: FBA = Farm Business Assets;
La = Land of Farm Business;
IC = Irrigation Capital;
FM = Farm Machinery for Farm Business;
OFA = Other Farm Assets.

Source: Estimated by authors.

With regard to HHA portfolio (Table 6.27), the dominant share is that of buildings, followed by HHD goods. Among farm groups, the share of building is higher in case of S and M farms and that of HHD goods higher in case of L farms.

Table 6.27

Disaggregative Analysis of Portfolio of Household Assets: All India: 2005-06

Percentage Share

Components	*S*	*M*	*L*	*Overall*
B	82	78	71	80
HHPROP	2	2	0	2
HHD	16	20	29	18
Total	100	100	100	100

Note: B = Buildings;
HHD = Household Durable Goods;
HHPROP = Other Property for Household.

Source: Estimated by authors.

Portfolio analysis of financial assets (Table 6.28) indicates that gold and deposits account for 77 per cent of total financial assets in overall, and the remaining share of 23 per cent is claimed by shares, currency and savings. In case of L farms, the share of deposits and shares is higher than in S farms.

Table 6.28

Disaggregative Analysis of Portfolio of Financial Assets: All India: 2005-06

Percentage Share

Financial Assets	*S*	*M*	*L*	*Overall*
Deposits	20	27	28	23
Savings	5	7	6	6
Shares	8	12	10	10
Gold	59	47	51	54
Currency	7	7	5	7
Total	100	100	100	100

Source: Estimated by authors.

In three agricultural productivity regions (Table 6.29), FBA forms over 80 per cent of total assets. In L farms of both MPR and HPR, NFBA claims 24 per cent and 28 per cent respectively of total assets.

Table 6.29

Composition of Assets: Three Agricultural Productivity Regions: 2005-06

Percentage Share

Assets	*LPR*				*MPR*				*HPR*			
	S	*M*	*L*	*Overall*	*S*	*M*	*L*	*Overall*	*S*	*M*	*L*	*Overall*
FBA	84	81	78	82	90	86	64	89	93	86	70	88
NFBA	4	3	0	3	1	4	24	2	1	10	28	7
HHA	10	13	19	12	8	9	11	8	5	3	2	4
LIVESTOCK	2	3	3	3	1	1	1	1	1	1	0	1
Total Assets	100	100	100	100	100	100	100	100	100	100	100	100

Note: LPR = Low-productivity states (MP, Rajasthan);
MPR = Medium-productivity states (UP, Bihar);
HPR = High-productivity states (Punjab, Tamil Nadu);
Other definitions as under Table 6.25;
Sample observations: S = 1435, M = 440, L = 30, Total obs. = 1905.

Source: Estimated by authors.

Further, in these farms of LPR and MPR, HHA claims the higher share in their respective total assets than in case of other two farm groups. Only in LPR, the share of livestock asset is 2 per cent and above in all farm groups.

It is observed from Table 6.30 that land capital among FBA forms the highest share of over 94 per cent in all the three regions in overall, and non-land capital assets form only 6 per cent or so. Next in importance to land is farm machinery and irrigation. The S farms of all three regions are starved of non-land capital stock, since the share of land capital is over 97 per cent. Yet other interesting result is that in large farms, the share of farm machinery is higher than in other two farm groups of the three regions.

Table 6.30

Disaggregative Analysis of Portfolio of Farm Business Assets: Three Agricultural Productivity Regions: 2005-06

Percentage Share

Components	*LPR*				*MPR*				*HPR*			
	S	*M*	*L*	*Overall*	*S*	*M*	*L*	*Overall*	*S*	*M*	*L*	*Overall*
La	97.1	92.1	80.2	94.4	98.5	95.8	91.0	97.8	98.8	98.3	98.5	98.6
IC	1.0	1.1	2.6	1.1	0.3	0.3	0.4	0.3	0.4	0.2	NA	0.3
FM	1.2	5.6	16.3	3.6	0.9	3.6	8.1	1.6	0.6	1.2	1.3	0.9
OFA	0.7	1.2	0.9	0.9	0.3	0.3	0.5	0.3	0.2	0.3	0.2	0.2
Total	100	100	100	100	100	100	100	100	100	100	100	100

Note: Definitions of abbreviations as under Tables 6.25 and 6.26.
Source: Estimated by authors.

The results from disaggregative analysis of HHA portfolio are given in Table 6.31. Building and HHD assets alone would form over 94 per cent of total HHA in all the three regions in overall. Building asset alone claims over 80 per cent of total HHA of S farms of all regions. The share of HHD assets in total HHA portfolio of L farms of all regions is higher than in case of S farms.

In all the regions, gold claims the highest share in financial asset portfolio of S and M farmers, followed by deposits and currency. But in case of L farmers of MPR, the highest share is claimed by deposits, followed by gold, and in case of L farms of HPR, the highest share is claimed by gold, followed by deposits (Table 6.32).

Table 6.31

Disaggregative Analysis of Portfolio of Household Assets: Three Agricultural Productivity Regions: 2005-06

Percentage Share

Components	*LPR*				*MPR*				*HPR*			
	S	*M*	*L*	*Overall*	*S*	*M*	*L*	*Overall*	*S*	*M*	*L*	*Overall*
B	84.2	78.9	81.7	81.7	80.8	76.3	81.3	79.7	82.4	77.8	63.6	79.8
HHPROP	0.1	0.6	NA	0.3	5.7	5.9	3.3	5.7	0.2	2.3	NA	1.0
HHD	15.7	20.5	18.3	18.0	13.5	17.8	15.4	14.6	17.4	19.9	36.4	19.2
Total	100	100	100	100	100	100	100	100	100	100	100	100

Note: Definitions of abbreviations as under Tables 6.25 and 6.27.
Source: Estimated by authors.

Table 6.32

Disaggregative Analysis of Portfolio of Financial Assets: Three Agricultural Productivity Regions: 2005-06

Percentage Share

Financial Assets	*LPR*				*MPR*				*HPR*			
	S	*M*	*L*	*Overall*	*S*	*M*	*L*	*Overall*	*S*	*M*	*L*	*Overall*
Deposits	14	25	27	20	25	41	55	32	11	23	39	18
Savings	1	2	3	2	1	2	2	1	2	5	3	3
Shares	1	0	5	1	4	5	4	4	1	2	0	1
Gold	78	63	59	69	62	45	36	56	76	60	54	69
Currency	6	10	6	8	8	7	3	7	10	10	4	9
Total	100	100	100	100	100	100	100	100	100	100	100	100

Source: Estimated by authors.

Determinants of Non-farm Business Assets

Dictated by data availability, it is postulated that demographic variables such as age of the head of the household, education status of the head and household size, along with household net worth and farm size are the determinants of NFBA accumulation. As evidenced in Table 6.33, only net worth and farm size have positive impact on NFBA accumulation, and all the demographic variables are found to have negative but statistically significant impact on dependent variables. Some of these demographic explanatory variables are found to have unexpected negative impact on dependent variables, compared to the results from NCAER 1998-99 data set.

Table 6.33

Bootstrap Tobit Regression of Non-farm Business Asset: All India (2005-06)

Variables	*Elasticities/Coefficients*	*Standard Error*	*t-value*	*P-value*
AoH	-3.55	1.60	-2.21	**
DEDU	-2.05	0.87	-2.36	**
NW	2.14	0.29	7.45	***
FS	6.64	1.30	5.13	***
HHS	-2.06	0.81	-2.55	***
Constant	-28.79	6.97	-4.13	***
Log Likelihood	-3336.82			
Number of Sample	3742			

Note: AoH = Log Function of Age of Head;
DEDU = Dummy variable of literacy/illiteracy of head where literacy = 1 and illiteracy = 0;
NW = Log Function of Net Worth (Net Worth = Total assets – Total outstanding debt);
FS = Log Function of Farm Size;
HHS = Log Function of Household Size;
'***' = Level of significance 1%;
'**' = Level of significance 5%;
'*' = Level of significance 10%.
Since almost more than 90 per cent observations of dependent variable are 0 and the rests are continuous, a censored model i.e., Tobit has been applied in the place of OLS with lower limit as 0. Bootstraps have been performed for each regression and they have been replicated 1000 times as done in case of Table 6.8.B. The rationale for using the technique of bootstrapping is under the section Determinants of Financial Inclusion/Exclusion.

Source: Estimated by authors.

Compositional Analysis of Liability Portfolio

Results in Tables 6.34-6.41 provide empirical insights into liability portfolio of sample households.

Overall, 66 per cent of total loan availed has come from institutional sources, and 34 per cent from non-institutional sources. The percentage share of loan availed by the three farm groups from institutional sources varies directly with farm size, and from non-institutional sources the percentage share varies inversely. In other words, large farmers are found to have availed 88 per cent of their outstanding loan from institutional sources, whereas it is 55 per cent in case of small farmers (Table 6.34).

Table 6.34

Compositions of Outstanding Loans: All India: 2005-06

Percentage Share

Components	*S*	*M*	*L*	*Overall*
IL	55	77	88	66
NIL	45	23	12	34
Total	100	100	100	100

Note: Sample observations: S = 2832, M = 862, L = 48, Total obs. = 3742.
Source: Estimated by authors.

In all the three regions, the large farmers have availed of more than 84 per cent of loan requirements from institutional sources, but small farmers only in HPR. For small farmers of LPR and MPR, non-institutional sources still continue to be dominant loan availment source (Table 6.35).

Table 6.35

Compositions of Outstanding Loans in Three Regions: 2005-06

Percentage Share

Financial Assets	*LPR*				*MPR*				*HPR*			
	S	*M*	*L*	*Overall*	*S*	*M*	*L*	*Overall*	*S*	*M*	*L*	*Overall*
IL	21	63	93	52	36	75	85	48	85	83	84	84
NIL	79	37	7	48	64	25	15	52	15	17	16	16
Total	100	100	100	100	100	100	100	100	100	100	100	100

Note: IL = Institutional loans;
NIL = Non-Institutional loans;
Sample observations: S = 1435, M = 440, L = 30, Total obs. = 1905.
Source: Estimated by authors.

Overall, the purpose-wise analysis of loan availment from institutional sources (Table 6.36) indicates that more than two-third of loan availed is for FB activities. Even in the case of small farmer, the percentage share comes to 59 for farm business, and the next highest is for acquisition of HHD assets. Added to this is the interesting result of small farmers availing 13 per cent for NFB activities, and this share is far higher than in the case of medium and large farmers.

Table 6.36

Component Analysis of Institutional Loan Availment: All India: 2005-06

Share of Total Outstanding Loan

Components	*S*	*M*	*L*	*Overall*
FB	59	77	69	68
NFB	13	8	2	10
HHD	15	7	12	11
Other	13	8	17	11
Total	100	100	100	100

Note: FB = Farm Business;
NFB = Non-farm Business;
HHD = Household Durables.

Source: Estimated by authors.

The purpose-wise analysis of loan availed by households from non-institutional sources (Table 6.37) suggests that miscellaneous purpose (including consumption, conducting of ceremonies and so on) dominates in overall, followed by loan availed for farm business activities and HHD asset acquisition. A higher percentage of loan availed is for 'other' purposes in case of small farm households, and it is for FB activities in case of L farm households.

Table 6.37

Component Analysis of Non-institutional Loan Availment: All India: 2005-06

Share of Total Outstanding Loan

Components	*S*	*M*	*L*	*Overall*
FB	24	38	54	28
NFB	6	3	0	5
HHD	22	20	12	22
Others	48	39	34	45
Total	100	100	100	100

Source: Estimated by authors.

In general, out of the total outstanding loan availed, irrespective of sources, the highest percentage (55%) is found to be for FB activities, followed by other purpose and acquisition of household durable assets. As between farm groups, both large and medium farmers have availed about two-third of loan for FB activities, and it is about 43 per cent in case of small farmers. But small farmers have availed a higher percentage of loans for NFB activities, for acquisition of HHD assets and other purposes than other two farm groups (Table 6.38).

Table 6.38

Component Analysis of Total Loan Availment: All India: 2005-06

Share of Total Outstanding Loan

Components	*S*	*M*	*L*	*Overall*
FB	43	68	67	55
NFB	10	7	2	8
HHD	18	10	12	15
Others	29	15	19	22
Total	100	100	100	100

Source: Estimated by authors.

Region-wise analysis of purpose-wise availment of loans could be examined from Tables 6.39-6.41.

- Cutting across farm groups, over 70 per cent of loan availed from institutional sources is for FB activities. In LPR, the next highest share is for other purposes, followed by acquisition of HHD assets. In MPR, the next highest is for other purposes followed by NFB activities. In HPR, the next highest is for NFB activities, followed by other purposes and acquisition of HHD assets.

  As between farm groups, both medium and large farmers have availed the highest percentage of loan for FB activities in all the regions, whereas small farmers have availed a greater percentage for NFB activities in both MPR and HPR regions than other two groups (Table 6.39).

- In case of loans availed from non-institutional sources (Table 6.40), in overall, the highest percentage of loan availed in LPR and MPR is for other purposes, followed by HHD, and in HPR the highest percentage is for FB activities. The small farmers of LPR and MPR have availed over 50 per cent of loans for other purposes. Large farmers in all the regions have availed a greater percentage of loan for farm business activities.

Table 6.39

Component Analysis of Institutional Loan Availment in Three Regions: 2005-06

Percentage Share in Total Outstanding Loan

Components	LPR				MPR				HPR			
	S	M	L	Overall	S	M	L	Overall	S	M	L	Overall
FB	64	81	60	74	66	84	100	75	60	80	61	71
NFB	1	0	5	1	11	10	0	10	29	0	0	12
HHD	15	9	5	9	4	4	0	4	6	8	39	8
Others	20	10	30	16	19	2	0	11	5	12	0	9
Total	100	100	100	100	100	100	100	100	100	100	100	100

Source: Estimated by authors.

Table 6.40

Component Analysis of Non-institutional Loan Availment in Three Regions: 2005-06

Percentage Share in Total Outstanding Loan

Components	LPR				MPR				HPR			
	S	M	L	Overall	S	M	L	Overall	S	M	L	Overall
FB	25	35	42	29	24	32	40	25	44	53	53	50
NFB	3	3	0	3	9	3	0	8	0	5	0	3
HHD	21	22	19	22	16	15	33	16	24	19	0	20
Others	51	40	39	46	51	50	27	51	29	23	47	26
Total	100	100	100	100	100	100	100	100	100	100	100	100

Source: Estimated by authors.

- Out of the total loan outstanding, in overall, the highest percentage is for FB activities, followed by other purposes, NFB activities and HHD (Table 6.41). Large farmers in all the regions have availed a high percentage of loan for FB activities. Small farmers of MPR and HPR have availed a higher percentage of loan for NFB activities than medium and large farmers of these regions.

Table 6.41

Component Analysis of Total Loan Availment in Three Regions: 2005-06

Percentage Share in Total Outstanding Loan

Components	*LPR*				*MPR*				*HPR*			
	S	*M*	*L*	*Overall*	*S*	*M*	*L*	*Overall*	*S*	*M*	*L*	*Overall*
FB	33	64	59	53	39	70	91	49	57	75	59	67
NFB	3	1	4	2	10	8	0	9	25	1	0	11
HHD	20	14	6	15	12	7	5	10	8	10	33	10
Others	44	21	31	31	40	14	4	32	9	14	8	12
Total	100	100	100	100	100	100	100	100	100	100	100	100

Source: Estimated by authors.

Relationship between Credit Propensity and Share of Assets: 2005-06

One of the important areas of relationship is the relation between credit availed per unit of area cultivated and percentage share of four types of assets. The results on this relation are derived from polynomial regression as shown in Figure 6.3.

- It is observed that percentage share of FBA is constant for S farms, and increasing for L farms with an increase in credit propensity. In case of M farms, the percentage share of FBA increases initially and then decreases with an increase in credit propensity.
- Percentage share of HHA decreases in case of S farms, for M farms the share increases at an increasing rate and in case of L farms the share decreases marginally.
- Percentage share of NFBA and livestock remains constant over credit propensity for M and L farms, and the share of NFBA decreases constantly in case of S farms.

The orders of percentage share of assets are almost similar for all the farm groups, where FBA share is the highest in the order followed by HHA, livestock and NFBA. The concluding observation for M farm is that as credit propensity increases, percentage share of FBA decreases followed by increment in percentage share of HHA. In case of L farms as credit propensity increases, percentage share of FBA increases, followed by declining share of HHA and livestock.

Figure 6.3

Relationship between Credit Propensity and Share of Assets: 2005-06

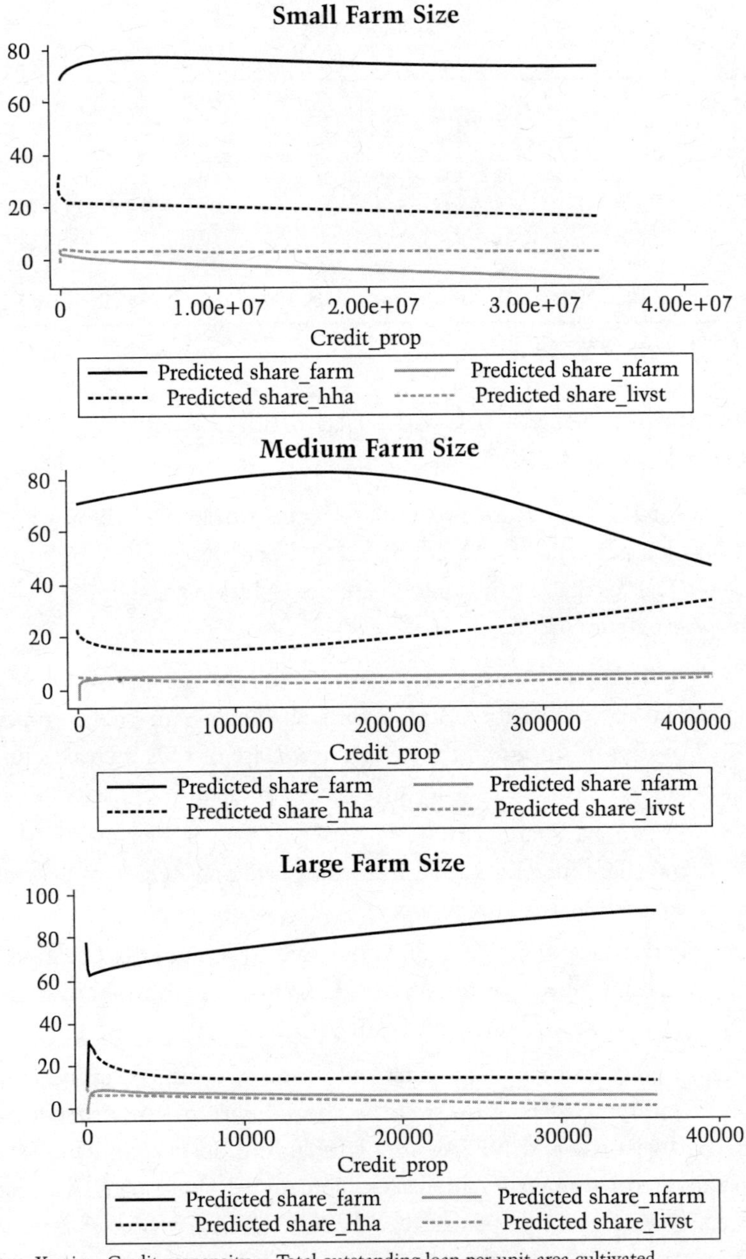

Note: X-axis = Credit propensity = Total outstanding loan per unit area cultivated.
Y-axis = Percentage share of each asset in total assets.

It is recalled from the discussion preceded that FBA commands higher share in the total value of assets. Hence the relationship between FBA, IL and NIL is displayed in Figure 6.4.

Figure 6.4

Relationship between FBA, IL and NIL in India: 2005-06

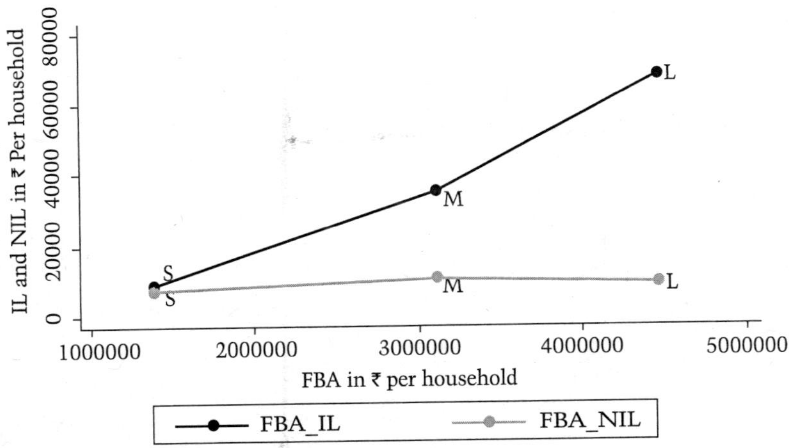

Note: X-axis = FBA in ₹ per household;
Y-axis = IL and NIL in ₹ per household.

- Black line (FBA_IL) shows the relationship between FBA and IL in 2005-06 for India. Grey line (FBA_NIL) shows the relationship between FBA and NIL in 2005-06 for India. S, M and L mean small, medium and large farms.
- As black line goes from S to M farms, FBA increases along with IL. As black line goes from M to L farms, FBA still increases along with IL and the intensity is higher than earlier.
- As grey line goes from S to M farms, FBA increases marginally along with NIL. As grey line goes from M to L farms, FBA decreases marginally along with NIL. Therefore when FBA increases, NIL remains almost constant across farms.
- When we compare both the black and grey lines together, we see that as FBA increases, IL increases and NIL remains constant.

Determinants of Financial Inclusion/Exclusion

It is postulated that education level of the head, age of the head, the level of FBA, NFBA, household assets, livestock and farm size are the key determinants of financial inclusion/exclusion. It could be observed from results in Table 6.42 that:

- With an increase in the age of household head, NFBA, HHA and farm size, the likelihood of financial inclusion rises. All these variables are statistically significant.
- But the surprising result is that with an increase in the education of the head of the household, the likelihood of financial inclusion falls, and the variable is statistically significant. Even though FBA and livestock asset variables are found to decrease the likelihood of financial inclusion, the variables are not statistically significant.

Table 6.42

Bootstrap Logistic Regression of Financial Inclusion: All India 2005-06

Variables	*Odds Ratio*	*Standard Error*	*t-value*	*P-value*
EoH	0.920931	0.013	-5.80	***
AoH	1.012684	0.003	4.48	***
FBA	0.977086	0.018	-1.24	
NFBA	1.075841	0.010	7.61	***
HHA	1.507659	0.071	8.72	***
Livestock	0.999628	0.009	-0.04	
FS	1.677122	0.140	6.20	***
Log Likelihood	-2261.01			
Number of Sample	3742			

Note: EoH = Education of Head;
AoH = Age of Head;
FBA = log of Farm Business Assets;
NFBA = log of Non-Farm Business Assets;
HHA = log of Household Assets;
Livestock = log of Livestock;
FS = Farm Size;
'***' = Level of significance 1%;
'**' = Level of significance 5%;
'*' = Level of significance 10%.

Source: Estimated by authors.

Capital Expenditure Pattern and Determinants

Capital expenditure incurred by sample households on four types of capital *viz.*, FBC, NFBC, HHC such as buildings and land, and on livestock capital during the year 2005-06 is analysed.

As shown in Table 6.43, more than 50 per cent of capital expenditure during 2005-06 is found to be towards household buildings and land (HHC). The next in the order of importance is capital expenditure towards farm business (FBC). With respect to capital expenditure, it is over 3 per cent in overall, in case of livestock, and it is quite insignificant in case of NFBC. With respect to FBC, the percentage share varies directly with farm size.

Table 6.43

Composition of Capital Expenditure: All India: 2005-06

Percentage Share

Components	*S*	*M*	*L*	*Overall*
FBC	36.6	43.7	46.5	39.7
NFBC	0.2	0.6	0.2	0.3
HHC	60.2	52.0	51.5	56.7
LivestockC	3.1	3.8	1.9	3.3
Total	100	100	100	100

Note: FBC = Capital Expenditure on Farm Business;
NFBC = Capital Expenditure on Non-Farm Business;
HHC = Capital Expenditure on Household Buildings and Land;
LivestockC = Capital Expenditure on Livestock;
Sample observations: S = 2832, M = 862, L = 48, Total obs. = 3742.

Source: Estimated by authors.

Since farm capital expenditure is the main component in respect of non-farm capital formation, a disaggregative analysis of percentage share of components of FBC is attempted as shown in Table 6.44: Percentage share of irrigation capital and farm machinery alone would form about over 90 per cent of total expenditure overall. The percentage share expenditure on land is considerable (19%) only in case of L farms. Hence capital expenditure on irrigation and farm machinery dominates in total capital expenditure incurred by sample farm households.

Table 6.44

Composition of Farm Capital Expenditure: All India: 2005-06

Components	*S*	*M*	*L*	*Overall*
La	7	7	19	7
IC	49	48	44	49
FM	42	42	36	42
OFA	2	3	2	2
Total	100	100	100	100

Note: La = Farm Capital Expenditure on Land;
IC = Farm Capital Expenditure on Irrigation Capital;
FM = Farm Capital Expenditure on Farm Machinery;
OFA = Farm Capital Expenditure on Others.

Source: Estimated by authors.

Region-wise and farm-size-wise analysis of capital expenditure pattern is yet another way of identifying difference in capital expenditure pattern of farmers in different regions and of farmers with different farm sizes. As shown in Table 6.45:

- The dominant items of capital expenditure in all the regions, cutting across farm groups, are FBC and HHC. Next in importance is capital expenditure on livestock. As between farm groups, in LPR all farm groups have incurred almost all their capital expenditure on FBC and HHC. The same inference holds good in case of MPR and HPR. In all the regions, livestock capital expenditure, even though small, by both small and medium farmers is perceptible.

The composition of farm capital expenditure incurred by three farm groups in LPR, MPR and HPR is given in Table 6.46. Capital expenditure on irrigation capital and farm machinery would form the dominant expenditure share of over/around 90 per cent. This is true with respect to all farm groups of all three regions. Farm group differences as with respect to individual components are also observed. For example, small farmers of all regions have incurred a greater proportion of their capital expenditure on irrigation, whereas large farmers on farm machinery. The pattern of capital expenditure incurred by medium farmers of three regions would also suggest the dominant share of irrigation.

Table 6.45

Compositions of Capital Expenditure in Three Regions: 2005-06

Percentage Share

Components	*LPR*				*MPR*				*HPR*			
	S	*M*	*L*	*Overall*	*S*	*M*	*L*	*Overall*	*S*	*M*	*L*	*Overall*
FBC	47.1	55.6	48.7	51.6	24.3	30.2	19.2	25.7	65.8	37.9	51.6	56.2
NFBC	0.3	0.2	0.0	0.2	0.1	0.0	0.0	0.1	0.0	0.0	0.0	0.0
HHC	48.9	42.1	49.6	45.5	71.3	64.8	80.8	69.8	32.8	57.9	47.8	41.6
Livestock	3.7	2.1	1.7	2.7	4.3	5.0	0.0	4.4	1.4	4.2	0.6	2.2
Total	100	100	100	100	100	100	100	100	100	100	100	100

Note: Definitions as under Table 6.42.
Sample observations: S = 1435, M = 440, L = 30, Total obs. = 1905.

Source: Estimated by authors.

Table 6.46

Compositions of Farm Capital Expenditure in Three Regions: 2005-06

Percentage Share

Components	*LPR*				*MPR*				*HPR*			
	S	*M*	*L*	*Overall*	*S*	*M*	*L*	*Overall*	*S*	*M*	*L*	*Overall*
La	3	3	5	3	13	15	11	13	1	2	1	1
IC	90	78	81	83	64	45	39	57	40	56	6	42
FM	6	16	14	12	22	40	50	29	58	40	88	55
OFA	1	3	0	2	1	0	0	1	1	2	5	2
Total	100	100	100	100	100	100	100	100	100	100	100	100

Note: Definitions as under Table 6.43.

Source: Estimated by authors.

The results on determinants of total capital expenditure are given in Table 6.47.

The determinants of capital expenditure are postulated to be total assets, total debt, household size, age of the household head, literacy and farm size. Excepting the age of the household head, all the other elasticities are found to be not only positive, but also statistically significant. With an increase in total assets, capital expenditure also increases. An increase in total debt in household size, in education of the household head and in farm size are all found to have increased total capital expenditure. The age of the household head, even though statistically insignificant, is found to have positive impact on capital expenditure.

Table 6.47

Bootstrap OLS Regression of Total Capital Expenditure: All India 2005-06

Variables	*Elasticities/Coefficients*	*Standard Error*	*t-value*	*P-value*
ASSET	0.455	0.030	14.93	***
DEBT	0.331	0.034	9.79	***
HHS	0.215	0.085	2.53	**
AoH	0.121	0.169	0.72	
DEDU	0.343	0.092	3.74	***
FS	0.575	0.141	4.08	***
Constant	-2.244	0.697	-3.22	***
Adjusted R Square	0.1499			
Number of Sample	3742			

Note: ASSET = Log Function of Total Assets;
DEBT = Log Function of Total Debt;
HHS = Log Function of Household Size;
AoH = Log Function of Age of Head;
DEDU = Dummy variable of literacy/illiteracy of head where literacy = 1 and illiteracy = 0;
FS = Log Function of Farm Size.

Source: Estimated by authors.

Major Inferences Derived

This section derives major inferences from NCAER samples of the two periods *viz.*, 1998-99 and 2005-06. Cutting across farm groups, two periods (since the interval between the first and second period is not that long) and regions, the inferences are derived as related to asset-liability portfolio pattern, financial inclusion and capital expenditure pattern as well as its determinants.

- FBA in total assets, land (over 90%) in total FBA, building and household durables in total household assets, and gold and deposits under financial assets claim the dominant share.
- There are indications of large farm households moving towards NFB activities.

- Barring some differences, the variables such as education and age of the household head, household size, net worth and farm size are found to have positive impact on NFBA accumulation.
- In case of liability portfolio: (a) over two-third of total outstanding loan is availed by sample farm households from institutional sources. But for L farm households, institutional source is found to be more dominant, and for S farm households non-institutional source is still important. (b) In case of purpose-wise availment of loans, FB activity and 'others' (miscellaneous purpose) are found to be more important in overall. But for small farmers, miscellaneous purpose of loan availment is quite high, whereas large farmers have availed loans mostly for FB activity.
- With respect to relationship between farmers, propensity to borrow and share of four assets, FBA is the highest in the order, followed by HHA, livestock and NFBA.
- The positive determinants of financial inclusion are found to be education and age of the household head, NFBA, HHA, livestock and farm size.
- In case of capital expenditure pattern: (a) the highest share, especially in case of S farm households in overall, is claimed by livestock. But in case of large farmers, the highest share is for FBC. (b) Irrigation capital and farm machinery in the category of FBC expenditure, and building under household capital expenditure claim the dominant share.
- The level of debt, household size, age and education of the household head, and farm size have had positive impact on capital expenditure incurred by sample households.

7 Major Inferences and Policy Directions for Asset Build-up

There are many studies which have analysed the structure, composition, growth and impact of agricultural capital stock in Indian agriculture, using both micro (farm level) and macro level data base. But the focus of all these studies has been on only one broad item of capital stock *viz.*, farm business assets (FBA) accumulated by cultivator households. But cultivator households are 'managers' of four types of assets *viz.*, FBA, non-farm business assets (NFBA), household durable (HHD) assets and financial assets. The focus of the present study is on some major dimensions of all these assets, keeping in view that development is a process through which households' welfare could be increased or vulnerabilities reduced through rasing asset accumulation equilibrium itself, by improving the returns to existing assets and/or by accessing acquisition of new assets. With the establishment of asset-income poverty linkage and thereby asset-social and economic well-being linkage, documenting and analysing the asset-liability portfolio of cultivator households have become of paramount importance. The present study is a major attempt, using farm-level data.

Objectives of the Study

The specific objectives of this study have been to analyse investment portfolio structure, composition and growth of assets, and determinants of accumulation of NFBA. The next broad objective has been to analyse household liability structure and financial inclusion and to explain variations in debt-asset ratio of cultivator households. The study has also analysed capital expenditure structure and its growth, and the determinants of capital expenditure incurred by cultivator households. All these objectives form the basis for analysing the National Sample Survey Organisation (NSSO) data for the years 1991-92 and 2002-03. But in case of data set from the National Council of Applied Economic Research (NCAER) for the years 1998-99 and 2005-06, some modifications have been made.

Data Base on Assets, Liabilities and Capital Expenditure

The objectives set out for this study are empirically analysed, using unit-level (farm level) data generated by NSSO as a part of its decennial, 'All India Debt and Investment Survey' (AIDIS) for the years 1991-92 and 2001-02. The total sample of cultivator households comes to 24,035 for the year 1991-92 and to 48,295 for the year 2001-02. This has facilited comparison of results on household asset (HHA) structure, household liability structure and also their related aspects, and capital expenditure incurred by these sample households. Further, the second data set is used, drawing unit-level (household) data collected by NCAER for the years 1998-99 and 2005-06. The major inferences and thereby policy directions from this data set are integrated, wherever relevant, with NSSO data based analysis. However, the focus is mainly on results derived from NSSO data analysis.

Household Asset Portfolio Analysis: Major Inferences and Policy Directions

This section recalls the major inferences (Chapters 3 and 4) on cultivators' asset portfolio and derive the relevant policy directions for 'influencing' the asset portfolio and its composition.

The aggregative results (cutting across farm size groups) lead to the inferences that:

- The empirical evidences establish the 'dominance' of FBA (forming two-third of the total value of assets), followed by HHA (forming about one-third), with NFBA and financial assets forming very insignificant share in total assets.
- Based on NCAER data analysis on asset portfolio during the two periods, it is inferred, cutting across regions and farm groups, that FBA in total assets, building and HHD in total HHA and gold and deposits under financial assets would form the dominant asset in terms of higher percentage share in respective total assets.

- There are two implications of this inference on 'dominant' asset pattern. First, the cultivator households are yet to diversify their asset portfolio, in terms of transition to accumulation of NFBA so as to facilitate pursuit of NFB activities to reduce the risks of income and consumption volatilities. The NCAER data analysis is suggestive of large farmers moving towards accumulation of NFBA. The vulnerability of cultivators' household livelihood security could be addressed by emphasing accumulation of NFBA and thereby encouraging NFB activities in rural areas. More on this later. Second, the monetisation of rural economy in terms of holding financial assets in the asset portfolio is yet to make a dent. This justifies the policy emphasis on social banking, banking the unbanked segments to broaden financial inclusion. Hence efforts are needed at policy-programme front to ensure diversification of assets in terms of accumulation of NFBA for NFB activities, and for participation of cultivator households in formal money and capital markets.

With respect to components of four broad groups of assets:

- Land forming over 90 per cent of FBA, animal capital forming two-third of NFBA, and residential building and land (for household activities) forming 86 per cent of portfolio of HHA provide needed directions for policies and programmes. First, the 'inadequacy' of non-land assets could be considered as a constraint to productivity of land and labour which have a direct bearing on prevalence of poverty and hunger among cultivator households. To ensure the accumulation of non-land capital assets, public investment policies and priorities become imperative (details in Bisaliah *et al.*, 2011). Some of the policy areas that need to be addressed are: public investment in irrigation, rural electrification, rural economic diversification, research and development to augment total factor productivity in agriculture, institutional innovations to forge the development alliance between small holders and corporate sector, rainfed farming and so on. Added to these is the need for public policy regime for inducing private investment (details in Bisaliah *et al.*, 2011). Second, the dominance of animal capital in components of NFBA provides insights into revealed preference of cultivator households in their asset accumulation behaviour. Policy support for accumulation and use of animal capital needs to be well placed, because the income elasticity of demand for animal product is quite high, in view of changing consumption pattern in both domestic and international markets.

The results derived with rural household samples also establish the dominance of FBA in total assets, of land in total FBA, of animal capital (forming 74%) in total NFBA of residential building and land (for household activities) forming about 87 per cent of total HHA.

- Among urban cultivator households, HHA and FBA would form 92 per cent of total assets, land forming about 94 per cent in FBA, animal capital and land (for non-farm business activities) forming over 50 per cent of total NFBA, and land and building (for household activities) would constitute 87 per cent of HHA. As between rural and urban cultivator households, there is more dominance of animal capital as a component of NFBA—as high as 74 per cent in rural households as against 29 per cent in urban households.

The major inference from regional analysis of asset structure is: in all the three regions of the study, FBA form two-third of total assets. Further, the dominance of land in FBA, of animal capital in NFBA and of residential building and land (for household activities) is established.

- In general, over a period of time, there has been trend towards decrease in the share of FBA in total assets, and marginal increase in the share of NFBA. Even though this is a 'desirable' transition to diversify rural economic activities, the trend has been slow.

Yet another way to examine the compositional shifts in asset portfolio is to examine growth rates of broad asset group. The highest growth rate of over 9 per cent is recorded in case of NFBA with implications for trend towards transition to NFB activities. Further, the farm households of low-productivity region (LPR) have experienced the highest growth rate of total assets, and the highest growth rate is observed much more so in case of growth rate of 10.6 per cent in NFBA of these households. In other words, the high-productivity region (HPR) is in the trajectory of low rate of growth in total assets, whereas the LPR is in the trajectory of high growth rate. It is too early to be conclusive about reduction in regional asset ownership disparities. But obviously change in 'asset ownership' could be one of the policy instruments to reduce regional disparities. Providing triggers for this could be a policy instrument of public investment.

The need for diversifying asset portfolio especially by 'encouraging' through policies and programmes, and the diversification towards more accumulation of NFBA has been well placed for rural economic transformation. As discussed earlier, there has been some empirical regularities to indicate positive relationship between accumulation of NFBA (thereby NFB activities) and income as well as household welfare. Infact, non-farm activity through the accumulation of non-farm assets offers a ladder out of poverty to cultivator households. This would warrant public policy support and public investment to facilitate the accumulation of NFBA and pursuit of NFB activities in rural areas. The term 'rural' is no longer to be treated synonymous with agriculture or food production system. China has demonstrated the evolution of new concept of 'rurality' within agricultural sector, and operationalised it with the support of public policies and investment for rural non-farm sector (RNFS). The RNFS in China (covering informal manufacturing, agro-processing, trade and so on) is closely connected to development of small towns. In fact, China has achieved rural economic diversification by supporting Township and Village Enterprises (TVEs). But all these would need 'big push' in public investment to support accumulation of more NFBA by Indian cultivator households and thereby to pursue NFB activities. But in India, non-farm investments have received little attention at policy level.

In the first place there is a need for identifying determinants of accumulation of NFBA by cultivator households. The present study has provided some empirical insights into these determinants. In fact, little is known about NFBA held by farmers, and the variables that influence the accumulation of these assets. The present study (cutting across rural and urban cultivator households, and farm groups) has established empirical evidences to indicate positive and significant impact of age of the head, net worth of the household, farm size and household size on accumulation of NFBA. NCAER data analysis also validates the postulate that education and age of the household head, household size, net worth and farm size have positive impact on NFBA accumulation. But policy and programme support is required through development education for credit availment, technical literacy (skill development) and commercial capacity of households to manage NFBA. The right institutional arrangement for providing 'literacy' in these three domains is to work with Special Help Groups (SHGs) to prepare these groups for group activities for accumulation of NFBA and for management of NFB activities. Even though this kind of transformation is a slow process, this is the best option for sustainability of this kind of activity among cultivator households.

Household Liability Analysis: Major Inferences and Policy Directions

Liability Composition

Cutting across rural and urban cultivators and farm groups, the share of institutional credit has decreased marginally and that of non-institutional sources increased during the second period. This inference is also validated by the results derived from NCAER 1998-99 and 2005-06 data set. This is also true in case of sample rural cultivator households, even though the share of institutional sources as high as 83 per cent during the first period has declined to 70 per cent during the second period. Region-wise analysis also leads to the inference that the percentage share of institutional loans to total outstanding loans incurred by cultivator households has declined over a period of time.

Declining share (however small) of debt from formal sources in total outstanding loan of cultivator households has sharpened the concepts of social banking, financial inclusion and banking the unbanked. In this context, the policy drive for financial inclusion of agricultural sector is well placed. But efforts in the direction of developing technological designs to 'transmit' financial services, and design of financial instruments to wean away farm households from investment in traditional assets to financial assets are urgently needed.

Loan Availment Components

Cultivator households have availed of loans for the purpose of farm business (FB), NFB, households, and a small portion for financial business, and other needs. The result with all-India sample leads to the inference that the percentage share of FB in total loans availed by cultivator households has increased by 12 percentage points between 1991-92 and 2002-03, and the share of all other components has declined. The same inference holds good in case of availment of loans by rural households, and it is true marginally in case of urban households also. Region wise, the pattern of availment is not anything different. Hence cultivator households tend to borrow a greater proportion for meeting the capital needs of FB activities. NCAER data analysis suggests three major inferences on purpose of loan availment. First, FB in general constitutes the major purpose, followed by loans availed for miscellaneous purpose (others) such as consumption needs and other social needs. Second, a greater percentage of loan availed by large farmers is for FB activities, and it is for other purposes (miscellaneous) in case of small farmers. Third, with respect to relationship between farmers propensity to borrow and share of four broad groups of assets, the highest share is claimed by FBA, followed by HHA, livestock and NFBA.

Debt-Asset Ratio: Major Inferences

Debt-intensity of farm households per unit of asset has increased over a period of time, and the institutional debt-intensity also has increased over a period of time. It is also true in case of farmers of LPR and medium-productivity region (MPR). In fact, in case of cultivators of LPR, the total debt-intensity is twice that of MPR and HPR.

Two implications could be derived for policy direction. First, the debt-intensity (i.e., debt per unit of asset) has increased over a period of time. Second, the farm households of LPR have incurred more debt per unit of asset, because of both low asset base and high propensity to borrow to meet their 'expenditure,' in view of low income from farm business and NFB activities. Hence policy directions are needed for increasing the asset base through the strategy of financial inclusion.

Determinants of Financial Inclusion: Major Inferences and Policy Directions

Ignoring inter-period minor differences, there is a higher probability that financial inclusion of an average rural farm household increases with an increase in age of the head, education status of the head, value of NFBA, farm size and FBA. The analysis performed with NCAER comes to the same conclusion.

In case of the three regions, in general the probability of financial inclusion increases with an increase in age of the head, farm size, education status, and the value of FBA and NFBA. Further, the probability of financial inclusion increases only in case of households of HPR, compared to those of LPR.

Three implications from the results on financial inclusion could be derived for policy direction. First, it is the value of farm business and NFBA which would increase the probability of financial inclusion for a farm household. In a formal credit system, where the 'insistence' is on security, although inevitable, the households with low asset base are prevented from getting access to credit from formal credit institutions, and are 'driven' to borrow from non-institutional sources. More of policy directions are needed to encourage 'collective security' system, instead of insisting on individual security. The 'collective security' system is already prevalent in case of credit for self-help groups (SHGs). Second, it is understandable that an increase in farm size increases the probability for financial inclusion, leading to the probability of exclusion of small holders. Third, special drive in policy front is needed to see that banks engage in proving more credit for building up non-farm assets and thereby credit worthiness of farm households. Fourth, the results on less probability of financial inclusion in case of farm households of LPR, even though understandable, need to examine whether region-specific financial inclusion policies are feasible.

Determinants of Debt-Asset Ratio: Major Inferences

Ignoring some variations in results derived from all-India rural and urban samples, and regional rural and urban samples and the two time periods, debt-asset ratios are found to have decreased with an increase in age of the head, education status of the head, household size, farm size and total capital expenditure. Almost the same inferences could be drawn with aggregative (total outstanding loan/total asset) analysis and with disaggregative (total outstanding farm business debt/total farm business asset).

These inferences, perhaps, could be postulated *a priori*. Because, the experience of the head, the education status of the head, farm size, household size (possibly with multiple sources of income) and total capital expenditure (a proxy for financial status of the household) could lead to decrease in debt per unit of asset owned by the household. Perhaps the same set of explanatory variables could have positive impact on asset accumulation. Similarly, the same set of variables could also influence variations in debt-burden. The major message for policy direction is to support farmers for asset build up so as to reduce the debt-burden.

Structure of Capital Expenditure, Growth and Determinants: Major Inferences

Capital Expenditure Pattern on Farm Business Assets

Farm machinery claims about one-third of the total expenditure incurred by farmers. The next highest share is claimed by land development (LD), animal capital (AC) and irrigation capital (IC). However there has been an increase in the share of AC and IC, and decrease in the share of other two forms of capital assets. Intertemporal shifts in expenditure share in favour of AC and IC are suggestive of farmers' priorities for accumulating assets, and hence policy support is needed for these components of FBA.

Region-wise AC claims the highest share in LPR, and both in HPR and MPR the highest share is in favour of farm machinery (FM). This allocation of capital expenditure sets the need for region-specific policy support.

Capital expenditure pattern as per NCAER data analysis suggests that: (a) highest share, in overall, is claimed by livestock. Highest share is claimed by farm business capital in case of large farmers. (b) The dominant share is claimed by irrigation and FM under FB capital expenditure, and under household capital expenditure the dominant share is claimed by building.

Capital Expenditure Pattern on Non-Farm Business Assets

The highest share of capital expenditure incurred is in favour of transport equipment, followed by building and machinery for NFB activities.

Region wise, in all the three regions, the capital expenditure share dominance (between 68% and 55%) is in favour of transport equipment.

The capital expenditure share dominance is in favour of transport equipment for NFB activities, suggesting this as a source of income.

Capital Expenditure Growth Rates

The highest growth rate (4.8%) is found in case of AC, followed by IC among components of farm capital expenditure at all-India level. With respect to region-wise growth rate of farm capital expenditure, in LPR the growth rates of only irrigation capital and of land development have been positive, but in case of AC and FM the growth rates have been negative. In MPR and HPR, the expenditure growth rates of all the farm capital components are positive and high.

In case of growth rates of expenditure of NFBA, the growth rate in total has been as high as 4.8 per cent, with high growth rates of capital expenditure on buildings, machinery and transport equipment for NFB activities.

Region wise, the overall growth rate of expenditure on non-farm assets is the highest in LPR, followed by MPR and HPR. Component wise, highest growth rate is found in case of building followed by machinery for NFB activities in LPR. In MPR, machinery and transport equipment have experienced high growth rates of over 12 per cent, whereas in HPR the growth rates are positive and high only in case of building and transport equipment for NFB activities.

What inferences could be extracted for policy directions. First, even though highest growth rate of AC is observed at all-India level, the growth rate is negative in LPR, but positive in case of growth rate of irrigation capital in this region. The negative growth rate of animal capital in a low agricultural productivity region provides the basis for policy direction to extend more support for the growth of AC. Second, the high growth rate of capital expenditure on non-farm business assets especially in LPR needs to be further supported from policy front to the development of NFB so as to diversify rural economic activities, a pattern that has emerged in China.

Asset Acquisition and Depletion Patterns

Cultivator households at all-India level have revealed their preference for acquiring more of FBA and HHA. In all the three regions, the same pattern of asset acquisition is observed.

With respect to depletion of assets also, a high percentage (95%) has taken place in FBA and HHA. Almost the same pattern of asset depletion has surfaced in all the three regions. If the high depletion of FBA is due to natural disasters, coping mechanism in the form of insurance is quite imperative for sustaining the FBA base for agriculture.

Capital Expenditure-Debt Ratio

The outstanding term loan/total capital expenditure analysis suggests that:

- At all-India level, farm households have greater amount of outstanding term loan for every unit of capital expenditure incurred on farm assets during the second period than during first period.
- Almost the same pattern is observed in all the three regions.

This may be suggestive of the farm households depending heavily on debt for meeting their capital expenditure.

Determinants of Total Capital Expenditure

In this study, it has been empirically established with rural sample that with an increase in total assets, in total debt, in household and farm size, in age of the head and in education status of the head, total capital expenditure on assets increases. Almost the same pattern of impact of explanatory variables on total capital expenditure on assets is observed in all the regions. It is obvious that farmers with more asset accumulation, better access to credit, high education status, larger family size (a proxy for multiple sources of income) and more experienced heads could afford to incur high level of capital expenditure on assets. In case of NCAER data analysis on determinants of capital expenditure, debt, household size, age and education of the household head, and farm size are found to have positive impact on capital expenditure incurred by sample households. Physical and human capital, and other family demographic characteristics could facilitate positive change in capital expenditure. But farmers with poor 'resource endowments' of all these are likely to be bypassed in building capital assets through capital expenditure.

Portfolio Analysis: Major Contrasting Inferences between Small and Large Farms

This section is an attempt to extract major inferences derived from the empirical analysis of asset portfolio, liability (debt) portfolio, and capital structure growth and determinants in case of only two contrasting farm size groups *viz.*, small and large. The results derived mainly from NSSO all-India sample are used for the purpose of contrast.

Asset Portfolio

Among four broad asset groups considered in this study, the share of FBA in total value of assets is as high as 79 per cent in case of large farms, and a little over 50 per cent in case of small holders.

In case of both farm groups, land alone contributes over 90 per cent to total FBA. This obviously establishes the 'inadequacy' of non-land capital assets, about which reference has already been made under the section 'Household Asset Potfolio Analysis: Major Inferences and Policy Direction.'

However, small over a period of time, there has been a marginal increase in the share of NFBA (in total assets) by two percentage points in case of small holders, and one percentage point in case of large farmers. This implies low asset diversification towards NFB activities, to promote which policy directions have been indicated under the above mentioned section of this chapter. The NCAER data analysis suggests that large farm households are moving towards NFB activities at a higher rate.

Further, among NFBA, it is animal capital that dominants over a period of time with an increase in share from 44 per cent to 68 per cent in case of small farms, and from 44 per cent to 46 per cent in case of large holdings. It is quite obvious that animal husbandry is the revealed preference of small holders for diversifying their source of income. The need for policy direction in this case has been suggested under the above mentioned section.

Even though asset base of small holder is low, the growth rates of both FBA and NFBA are found to be higher in case of small holders than large farmers. In fact, the same pattern of growth is observed in all the three regions.

However, the regression results on determinants of NFBA accumulation derived with all-India rural sample suggest that with an increase in farm size, accumulation of NFBA also increases, and decrease in farm size has negative impact on NFBA accumulation. This leads to the development message, small holders need a little more policy and programme support for asset diversification especially with more of NFBA for activity and income diversification.

Debt Portfolio

The analysis performed with all-India sample suggests that large farmers have increased their loan share from institutional sources over a period of time, implying trend towards more of financial inclusion of these farm groups. On the other hand, the small holders have experienced a fall in the share of institutional loans in the total outstanding loans, implying trend towards more of financial exclusion. Almost the same inference holds good in case of cultivator households of three regions. The NCAER data analysis validates the inference of institutional source as a dominant source of loan for large farmers, and that non-institutional source still continues to be an important source for small farm households. This has been further reinforced by the logistic regression analysis on financial inclusion. As per the results derived from this analysis, there is a higher probability of financial inclusion with an increase in farm size. This result also implies that the probability of 'financial inclusion' decreases with a decrease in farm size. It has already been discussed under the section 'Determinants of Financial Inclusion: Major Inferences and Policy Direction' about kind of policy directions needed to address this issue. There is a higher probability of financial inclusion with an increase in FBA and NFBA. This also implies that there is low probability of small holders getting access to institutional credit, in view of their low asset base.

Over a period of time, in case of both small and large farmers, there has been an increase in the percentage share of loan (in total loan availed) availed for farm business, and a fall in the share of NFB, household durables and others. As per NCAER data based results, large farmers have availed loans mostly for farm business activity, and small farmers for miscellaneous purpose. But unless efforts are made in policy front to increase the share of NFB, especially in case of small holders, the needed asset-activity-income diversification will not occur to reduce the income and consumption risks of rural farm households.

Debt-asset ratios over a period of time for both small and large farms have turned out to be the same. With lower asset base, small holders tend to be adversely affected in terms of loan availment per unit of asset. They cannot get access to more credit needed to build up their asset base.

Capital Expenditure Analysis

In case of small farmers, capital expenditure on AC, LD, and FM would form 79 per cent of total capital expenditure, under capital expenditure on FBA. Whereas in case of large farmers, FM, LD and IC claim 80 per cent of total capital expenditure.

When it comes to composition of capital expenditure on NFBA, expenditure on transport equipment is the most dominant one in case of small farmers, and in case of large farmers, capital expenditure on machinery is the most dominant asset.

With respect to growth rates of capital expenditure, growth rates of expenditure on components of FBA is both positive and high, but in case of large farms, growth rates have been negative.

In case of components of non-farm capital expenditure, small farms have experienced the highest growth rate of capital expenditure on building followed by transport equipment and machinery—all for NFB activity. Large farms also have experienced highest growth rate of expenditure on building, followed by machinery and transport equipment.

There is not much of a difference in the asset acquisition behaviour of small and large farmers. Both the groups have placed their asset acquisition priority on FB and HA. Almost the same pattern is observed in case of depletion of assets (dominance of FB and HHA) of both large and small farmers.

Outstanding term loan per unit of capital expenditure is higher in case of large farmers. This implies that the ability of large farmers to get access to term loan is perhaps higher. This obviously would support asset build up in case of large farms, even by resorting to borrowing.

The regression results on determinants of capital expenditure have established, among other variable, positive relationship between value of assets and capital expenditure, and farm size and capital expenditure. This obviously implies 'low ability' of small holders with low asset base and smaller holdings in increasing capital expenditure. This result reinforces the needed support through institutional credit system to help the small holders in their efforts to incur capital expenditure so as to build up asset, income, and sustainable livelihood base.

The present study has provided empirical evidences for policy, programme and implementation drive for providing development support to small holder, in view of their numerical dominance (about 80% of total holdings) in Indian agriculture, and for virtuous broad based growth from the angle of both equity and growth. These empirical evidences are reflected in: low base of non-land capital assets, insignificant share of NFBA in total assets, the revealed preference for AC asset, fall in the share of institutional loans out of total outstanding loan, small farm size experiencing decrease in the probability of financial inclusion, fall in the share of loan availed for the acquisition of NFBA with negative impact on asset-activity-income diversification matrix, and low ability of small holders to build up asset base through capital expenditure.

References

Adato, Michelle, Michael Carter and Julian May (2006). "Exploring Poverty Traps and Social Exclusion in South Africa Using Qualititative and Quantitative Data", *Journal of Development Studies* 42(2).

Andrew, Dorward, S. Anderson, S. Clark, B. Keane and J. Moguel (2001). *Asset Functions and Livelihood Strategies: A Framework for Pro-Poor Analysis, Policy and Practice*. Wye, U.K.: Agroecology Group, Imperial College.

Barret Christopher B., Thomas Reardon and Patrick Webb (n.d.) *Non-Farm Income Diversification and Household Livelihood Strategies in Rural Africa: Concepts, Dynamics, and Policy Implications*. Ithaka, N.Y.: Department of Applied Economics and Management. Cornell University,

Birthal, Petal (2012). "Livestock for Higher, Sustainable and Inclusive Agriculture Growth", *Economic and Political Weekly*, June 30.

Bisaliah, S. (2010a). *Capital Formation, Agriculture Growth and Poverty: Conceptual and Empirical Constructs*. Consultancy Report. Rome: FAO.

———. (2010b). "Capital Formation in Indian Agriculture: Major Issues and Policy Directions", in *Agriculture, Food Security, and Rural Development*. Manila: ADB and New York: Oxford University Press.

———. (2012). *Investment in Agriculture: Cross Country Evidences, Experiences and Business Models*. Consultancy Report. Rome: FAO.

Bisaliah, S. and S. Mahendra Dev (2010). *Private Capital Formation In Indian Agriculture: An Analysis of Farm Level Data*. Consultancy Report. Rome: FAO.

Bisaliah, S., S. Mahendra Dev and Syed Saifullah (2011). *Investment in Indian Agriculture: Macro and Micro Evidences*. Consultancy Report. Rome: FAO.

Carter, M.R. and Julian May (2001). "One Kind of Freedom: Poverty Dynamics in Post-Apartheid South Africa", *World Development* 29, November 12.

Chand, R., Praduman Kumar and Sant Kumar (2011). "Total Factor Productivity and Contribution of Research Investment to Agriculture Growth in India", *Policy Paper* 25. New Delhi: National Centre for Agricultural Economics and Policy Research.

Christoper, B.B., Ernest Aryeetey, Agnes Quisumbing, Akhter Ahmed, John Hoddinott, Felix Naschold, Jecqueline Vanderpuye-Orgle and Tassew Woldehanna (2007). "Local Risk Management: Protecting Household Asset Building and Farm Productivity from Idiosyncratic Shocks", *Basis Brief*. University of Wisconsin–Madison July.

Cramon-Taubadel, Stephan von, Gustavo Anriquez, Hartwig de Haen and Oleg Nivyevskiy (2009). "Investment in Developing Countries Food and Agriculture", *Expert Meeting on How to Feed the World in 2050*. Rome: Economic and Social Development Division, FAO.

Cobb-Clark, Deborah A., and Vincent A. Hildebrand (2008). *The Asset Portfolio of Native-Born and Foreign-Born Households*. Bonn: The Institute for the Study of Labour.

Deininger, K. *et al.* (2009). "Land Reforms, Poverty Reduction, and Economic Growth: Evidence from India", *The Journal of Development Studies*. March 25.

Deininger, Klaus, Songqing Jin and Hari K. Nagarajan (2011). "Can Political Reservations Empower Women and Affect Economic Outcomes: Evidence from Rural India", *NCAER Working Papers on Decentralisation and Rural Governance in India*. October 21. New Delhi.

Dev, S.M. (1998). *Inclusive Growth In India: Agriculture, Poverty and Human Development*. Oxford University Press, New Delhi.

Dev, S. Mahendra (2006). "Financial Inclusion: Issues and Challenges", *Economic and Political Weekly*, October 14.

Djankov, Simeon, Pedro Miranda, Enrique Seira and Siddharth Sharma (2008). "Who are Unbanked", *World Bank/International Finance Corporation Seminar*. February.

Ellis, F. (2000). *Rural Livelihoods and Diversity in Developing Countries*. Oxford University Press.

Gollier, C. (2000). *What Does Classical Theory Have to Say About Household Portfolios?* Universite de Toulouse and Institute Universitaire de Frame.

Government of India (2008). *Report of the Committee on Financial Inclusion*. Rangarajan Committee Report.

Guiso, Luigi and Tullio Jappelli (2000). *Household Portfolio In Italy*. Centre for Studies in Economics and Finance. University of Sassari, Italy.

Gulati, Ashok and Seema Bathla (2002). "Capital Formation in Indian Agriculture: Trends, Composition and Implications for Growth", *Occasional Paper* 24. Mumbai: National Bank for Agriculture and Rural Development.

Hanohan, P. (2004). "Financial Development, Growth and Poverty: How Close are the Links?", in G. Charles (ed.), *Financial Development and Economic Growth: Explaining the Links*. London: Palgrave.

Haveman, R. and Edward Wolff (2004). "The Concept and Measurement of Asset Poverty: Levels, Trends and Composition for the U.S: 1983-2001", *Journal of Economic Inequality* 2(2): 145-69, August.

Jain, Tarun (2011). *Where There is a Will: Fertility Behaviour and Sex Bias in Large Families*. Indian School of Business. October, 19.

Koril, K.L., K. Lagat Job and K. Njehia Benard (2011). "The Role of Non-Farm Investments in Agricultural Risk Management in Kenya", *Current Research Journal of Economic Theory* 3(2): 62-68.

Korir L.K. (2011). *Risk Management among Agricultural Households and The Role of Off-Farm Investments in VASIN GISHU County*. Kenya: Egerton University.

Levine, R. (2005). "Finance and Growth", in Phillipe Aghion and Steven N. Durlauf (eds.), *Hand Book of Economic Growth*. North-Holland.

Madhur *et al.* (2011). *Accelerating Agricultural Productivity Growth in India Policy and Investment Options*. Concept Note.

Manojit, B. and Meenakshi Rajeev (2010). "Interest Rate Formation in Informal Credit Markets in India: Does Level of Development Matter?", *BWPI Working Paper* 126. Books World Poverty Institute, The University of Manchester.

Michael, H. and Alexander Michaelides (2000). "Calibration and Computation of Household Portfolio Models", in Luigi Guiso, Michael Haliassos and Tullio Jappelli (eds.), *Household Portfolio*. MIT Press.

Mohan, R. (2008). "The Growth Record of Indian Economy: A Story of Sustained Saving and Investment", *Reserve Bank of India Bulletin*, March.

Moser, C. (2005). "Assets, Livelihood and Social Policy", Arusha Conference *New Frontiers of Social Policy*. December 12–15.

Moser, C.O.N (1990). *Reassessing Urban Poverty Reduction Strategies: The Asset-Vulnerability Framework*. World Bank Report.

Naschold, F. (2009). *Poor Stays Poor: Household Asset Poverty Traps in Rural Semi-Arid India*. Ithaca, N.Y.: Cornell University

NCAER (1975). *Changes in Rural Income in India*. New Delhi.

———. (1986). *Changes in Household Income, Inter-Class Mobility and Income Distribution in Rural India: A Longitudinal Study: 1971/72–1981/82*. New Delhi.

———. (1987). *Changes in Structure of Household Income and Distribution of Gains in Rural Household Sector: An All India Temporal Analysis 1971/72–1981/82*. New Delhi.

———. (n.d.). *Additional Rural Incomes Survey: Reference Period 1971 to 2006*. New Delhi.

NSSO (National Samply Survey Organisation) (1998). *Reports on 1991-92 Survey*. Government of India.

———. (2007). *Reports on 2001-02 Survey*. Government of India.

Pallavi, C. (2005). "How Inclusive are Banks under Financial Liberalisation", *Economic and Political Weekly*, October 23.

Prasanna, Chandra (2004). *Investment Analysis and Portfolio Management*. New Delhi: Mc Graw Hill.

Rajalaxmi, K., Arnab Mukherji and Maria Sandstrom (2010). "Accessing Institutional Finance: A Demand Side Story of Rural India", *Economic and Political Weekly* XLV(37), September 11.

Rajan, Raghuram and Luigi Zingales (1998). "Financial Development and Growth", *American Economic Review* 88(3).

Rao, S. (2007). "Financial Inclusion: An Introspection", *Economic and Political Weekly* February 3.

Serra, Teresa, Barry K. Goodwin and Allen M. Featherstone (2004). "Determinants of Investments in Non-Farm Assets by Farm Households", Paper for Presentation at the *American Agricultural Economics Association Annual Meeting*. Denver. July 1–4.

Thorat, U. (2006). "Financial Inclusion and the Millennium Development Goal", *Reserve Bank of India Bulletin*, February.

Vashishtha, P. (n.d.). *Changes in Structure of Investment in Rural Households: 1970/71 to 1981/82*. New Delhi: NCAER.

Vijay, R. (2012). "Structural Retrogression and Rise of 'New Land Lords' in Indian Agriculture: An Empirical Exercise", *Economic and Political Weekly*, February 4.